I
LJS16
31/3/80

3.95

574.192
085

KV-681-773

THE OPEN UNIVERSITY

Science : A Third Level Course

BIOCHEMISTRY AND MOLECULAR BIOLOGY

**BLOCK A
MACROMOLECULES**

INTRODUCTION TO BLOCK A :
Why the molecular approach ?

UNIT 1
INTRODUCTION TO MACROMOLECULES

I Introducing macromolecules
II Size and shape of macromolecules

UNIT 2
CONFORMATION

I Principles of conformation
II Experimental techniques

PREPARED BY AN OPEN UNIVERSITY COURSE TEAM

THE OPEN UNIVERSITY PRESS

The S322 Course Team

Chairman and General Editor

Irene Ridge

Authors

Norman Cohen
Bob Cordell (*Staff Tutor*)
Vic Daniels
Gerald Elliott
Anna Furth
Lindsay Haddon (*Course Assistant*)
Altheia Jones-Lecointe
Irene Ridge
Colin Self

Editor

Janet Evans

Other Members

Andrew Crilly (*BBC*)
Beryl Crooks (*IET*)
Vic Finlayson (*Staff Tutor*)
Denis Gartside (*BBC*)
Roger Jones (*BBC*)
Jack Koumi (*BBC*)
Jean Nunn (*BBC*)
Steven Rose
Don Wilson (*Senior Counsellor*)

Consultant for Block A

Professor C. F. Phelps (*University of Lancaster*)

The Open University Press,
Walton Hall, Milton Keynes.

First published 1977.

Copyright © 1977 The Open University.

All rights reserved. No part of this work may be reproduced in any form, by mimeograph or any other means, without permission in writing from the publisher.

Designed by the Media Development Group of the Open University.

Printed in Great Britain by
Martin Cadbury, a member company of Blackwell Press,
Worcester and London.

ISBN 0 335 04410 7

This text forms part of an Open University Course. The complete list of Units in the Course is printed at the end of this text.

For general availability of supporting material referred to in this text please write to the Director of Marketing, The Open University, P.O. Box 81, Walton Hall, Milton Keynes, MK7 6AT.

Further information on Open University Courses may be obtained from the Admissions Office, The Open University, P.O. Box 48, Walton Hall, Milton Keynes, MK7 6AB.

1.1.

Christ's and Notre Dame College
LIBRARY

Accession No. ND 56786

Class No. QUARTO
574.193
UNI

Catal. 22.5.84
AB.

L. I. H. E.
THE MARKLAND
LIBRARY

CLASS: 502
JO
ACCESS: 56786
CATL:

S322 Course Cover

The design is based on a low resolution model of haemoglobin

Introduction to Block A

Why the molecular approach?

LIVERPOOL INSTITUTE OF HIGHER EDUCATION
THE MARKLAND LIBRARY

You may already be familiar with some of the properties of macromolecules from earlier Courses. It is no exaggeration to say that the unique properties of living as opposed to non-living organisms can in all cases be traced back to the properties of their macromolecules. One reason why life has not yet been created in a test tube, is that the chemistry of macromolecules is much more complex than that of small molecules. Nevertheless, macromolecules do obey the same laws of chemistry and physics as do small molecules, and there is no need to postulate an extra ingredient of 'vitalism' as was done in the 19th century.

Bearing in mind that most biochemists are really interested in macromolecules only for what they can learn about the whole organism, or at least the whole cell, we shall try not to become lost in the finer points of chemistry or physics. However, there is no escaping the fact that recently most of the major advances in our understanding of biology have come from studies at the molecular level and any serious student of biochemistry must begin here. Although the trend may eventually be reversed as we reassemble isolated molecules back into subcellular components (e.g. membranes), it is essential to tackle the subject first at the molecular level. For this reason the first four Units of this Course are devoted to a study of macromolecules.

All four classes of biochemical compounds are discussed—lipids, polysaccharides, proteins and nucleic acids. The lipids, to which we shall return in Unit 3, do not fit easily into any general definition, but the polysaccharides, proteins and nucleic acids may all be thought of as polymers derived from the condensation of large numbers of component building blocks. In Unit 1 we shall consider what properties the macromolecules have in common, and how they differ from small molecules. There will be particular emphasis on the non-covalent bonding which is responsible for secondary, tertiary and quaternary structure. We shall then go on to consider ways of determining broad outlines of shape, i.e. whether the molecule is fibrous (long and thin) or globular (approximately spherical), and how the molecular weight may be determined. Unit 2 describes methods for investigating macromolecular shape in finer detail—fine enough to detect the small changes in conformation† that occur, for example, when an enzyme interacts with its substrate.

Most of the examples in Units 1 and 2 are drawn from proteins or nucleic acids. This is not to say that the polysaccharides or lipids are inherently less interesting, but until recently we have known relatively little about them. Unit 3 describes some of the principles that are beginning to emerge from recent work on these two classes of macromolecules.

Finally Unit 4 is a Case Study of one particular macromolecule, the enzyme ribonuclease, which is a protein that degrades nucleic acids. The Unit shows how information on the types of structure described in earlier Units can be built up to give a picture of how a macromolecule functions.

† Terms indicated thus (†) appear in the index of the S322 *Source Book for Biochemistry and Molecular Biology*, which should be consulted if you require further information.

Stereoslides for Block A

1 Carboxypeptidase, showing the position of active site tyrosine in absence of substrate

2 Carboxypeptidase, showing the position of active site tyrosine in presence of substrate

3 β-Pleated sheet

4 α-Helix

5 α-Helix and stabilizing H bonds

6 Lysozyme, active site

7 Sausage model of lysozyme

8 B form of DNA

9 RNA

10 Electron density map of lysozyme

11 Ribonuclease

12 Glyceraldehyde-3-phosphate dehydrogenase

UNIT 1 INTRODUCTION TO MACROMOLECULES

Contents

Table A1

Assumed knowledge

Principal Sections of prerequisite Courses, set books and of the *Source Book*,[a] which you may use in conjunction with Unit 1 pre-Unit tests. Much of the chemical information in these references is only a summary. For a fuller explanation you may like to look up the same terms in a standard chemistry textbook (see Recommended reading). Topics marked with an asterisk (*) are needed for understanding central concepts in the Unit, rather than for background examples.

Topic	Reference
absorbance, absorption maximum (λ_{max})	*Source Book*
allosterism	S2–1,[b] Unit 4, Section 4.2.3; *Source Book*, under enzymic catalysis
*covalent and ionic bonding	S100,[c] Unit 10, Section 10.3; *Source Book*
dissociation	*Source Book*
*electronegativity	S100, Unit 10, Section 10.1; *Source Book*
electrophoresis and gel filtration	S2–1, Unit 1, Section 1.2; *Source Book*
enzymic catalysis	S2–1, Unit 2, Section 2.3; *Source Book*
*α-helix, β-pleated sheet	S2–1, Unit 1, Section 1.5.2.
microscopy	S100, Unit 2
mitochondria	S2–1, Unit 3, Section 3.7.2.
optical isomers	S100, Unit 10, Section 10.4.5; *Source Book*
permeases	S2–1, Unit 5, Section 5.4; *Source Book*
*pH, pK	S2–1, Unit 2, Appendix 1; *Source Book*
*polar and non-polar molecules	S100, Unit 10, Section 10.1; *Source Book*
protein synthesis	S2–1, Unit 6, Sections 6.3, 6.4; *Source Book*
refractive index	S100, Unit 22, Section 22.4.1.
ribosome structure	*Cell Structure and Function*,[d] p. 279; *Source Book*
*rotation about double and single covalent bonds	S100, Unit 10, Section 10.4.3; *Source Book*, under covalent bonding

a The Open University (1977) S322 *Source Book for Biochemistry and Molecular Biology*, The Open University Press.

b The Open University (1972) S2–1 *Biochemistry*, The Open University Press.

c The Open University (1971) S100 *Science: A Foundation Course*, The Open University Press.

d A. G. Loewy and P. Siekevitz (1969) *Cell Structure and Function*, 2nd edn, Holt, Rinehart and Winston; the set book for S2–1.

Table A2

List of scientific terms and concepts in Unit 1

	Page No.		Page No.
absolute methods	30	molecular weight determination	30
autoradiography	45	monomer	11
		multimeric complex	16
bond energy	17		
building block	11	non-ideality	27
buoyant density	37		
		obsolescence	23
conformation	22		
		polyacrylamide gel electrophoresis, PAGE	32
denaturation	23	polymerization	11
density gradient centrifugation	35	primary structure	14
diffusion coefficient	28	pyruvate dehydrogenase complex, PDC	39
directionality	17		
		quaternary structure	15
electrophoretic mobility	32		
empirical methods	30	radius of gyration	28
		random coil	28
fibrous protein	13		
flexibility	22	secondary structure	14
fragility	23	sedimentation coefficient	35
frictional coefficient	28	sedimentation equilibrium	37
		sedimentation velocity	33
gel filtration	30	shapes of macromolecules	28
globular protein	13	sodium dodecyl sulphate, SDS	32
		specificity	21
heterogeneity	13	structural hierarchy	14
homogeneity	25	structural/storage macromolecules	13
hydrogen bond	17	subunit	15
hydrophobic bond	19	Svedberg unit	35
informational macromolecule	13	tertiary structure	14
intrinsic viscosity	29		
ionic bond	19	ultracentrifuge	33
		urea	31
light scattering	32		
		van der Waals bonding	19
macromolecule	11	viscosity	29
microenvironment	21		
molecular excluded volume	27	weak bonding	17

Objectives

After completing Unit 1 you should be able to:

1   Describe, by writing down chemical formulae or by using the model building kit, how macromolecules are built up from their constituent building blocks. (SAQs 1 and 2)

2   Distinguish between informational and structural/storage roles in macromolecules. (SAQ 3)

3   Predict the effect of bond strength, distance and (in the case of ionic bonds) pH, on macromolecular interactions. (SAQ 4)

4   Relate the properties of non-covalent bonds to the flexibility, fragility and specificity shown by macromolecules. (SAQs 7 and 8)

5   Predict from its polar or non-polar nature, whether a given residue in a macromolecule is likely to form van der Waals bonds rather than hydrogen or ionic bonds. (SAQs 5 and 6)

6   Predict the effect of changes in size, shape and flexibility on the properties of macromolecules in solution. (SAQs 5, 9 and 10)

7   Suggest techniques suitable for elucidating size and shape in different macromolecules. (SAQs 10 and 14)

8   Give examples to illustrate how biophysical techniques have widened our knowledge of biology. (SAQs 11, 12 and 13)

Pre-Unit test

This test should be completed before you proceed with the Unit. Answers are given on p. 51. If you have trouble in answering any of the questions you should look up the relevant Sections in other Courses, set books or in the *Source Book*, as indicated in Table A1. Questions 1–12 relate mainly to chemistry, questions 13–18 to biology.

1   Carbon forms stable bonds with other atoms by increasing the number of electrons in its outer shell from 4 to 8, in order to reach the stable electronic configuration of the nearest inert gas in the Periodic Table. True or false?

2   The difference between covalent and electrovalent (or ionic) bonds is that the shared electrons remain associated with both combining nuclei in *electrovalent (or ionic) bonds*, but are entirely given over to one or other nucleus in *covalent bonds*. True or false?

3   A covalently bound compound will be partially ionic or *polar* if the combining atoms differ in electronegativity. The electronegativities of fluorine, oxygen, chlorine, carbon and hydrogen atoms are: F, 4.0; O, 3.5; Cl, 3.0; C, 2.5; H, 2.1. Given this information, arrange the following compounds in order of decreasing polarity: H_2O, HCl, CO_2, HF. (You may find it helpful to indicate first the direction of polarization, as in $\overset{\frown}{H-Cl}$.)

4   Name the small molecules whose polymerization gives rise to each of the following macromolecules: (i) DNA, (ii) starch, (iii) a protein. *Note* Each of the macromolecules is a *polymer* formed from a string of identical or closely similar molecules, all covalently bonded together.

5   Calculate the pH of the following solutions:
  (i)   pure water (in which $[H^+]$ is only 10^{-7} mol l^{-1});
 (ii)   0.1 M HCl;
(iii)   0.01 M HCl.

6   An acid may be defined as any compound that dissociates in water to give free H^+ ions. Write down the equation for the dissociation of acetic acid (CH_3COOH) in water, and then rearrange this to give K, the equilibrium constant for the dissociation reaction.

7   Acetic acid, unlike say HCl, is a weak acid because at equilibrium the dissociation reaction is not fully displaced to the right. Given this information, and the definition of pK in the *Source Book*, indicate whether the following statements are true or false:
 (i)   A strong acid is one with a high K value.
(ii)   An acid of pK2 is stronger than one of pK7.

8   Using the relationship between pH and pK described in the *Source Book* (equation 6, under pK) answer the following questions:
(i)   The pK value of an acid is given by the pH at which the acid is exactly 50 per cent dissociated. True or false?
(ii)   Is an acid mainly in its undissociated form above or below its pK value?
(iii)   The imidazole ring in the amino acid histidine can ionize as shown in equation 1.1. If the pK is 6, what will be the charge on the ring at pH 7?

$$\begin{matrix} HC-\overset{H}{\underset{}{N^+}} \\ \parallel \quad \diagdown \\ -C-N \\ \quad H \end{matrix} \rightleftharpoons \begin{matrix} HC-N \\ \parallel \quad \diagdown \\ -C-N \\ \quad H \end{matrix} + H^+ \qquad (1.1)$$

9   Arrange the following compounds in order of increasing rigidity:

$$CHCl=CH_2, \quad CH_3-CH_3, \quad CH\equiv CH.$$

10   D-Lactic acid and L-lactic acid are mirror images of one another. True or false?

11   The absorption spectra of compounds A and B have the following characteristics:

$$\lambda_{max} = 340 \text{ nm} \qquad \epsilon = 20.0 \qquad \text{(compound A)}$$
$$\lambda_{max} = 280 \text{ nm} \qquad \epsilon = 1.0 \qquad \text{(compound B)}$$

(i) What will be the most sensitive wavelength at which to measure concentrations of compound A in solution?

(ii) Which compound absorbs most strongly (at its own λ_{max}), on a mole for mole basis?

12 In electrophoresis, molecules separate primarily because of differences in charge, while in gel filtration the separation is mainly due to differences in size. True or false?

13 A proteolytic enzyme will not hydrolyse carbohydrates, mainly because carbohydrates are the wrong shape to fit into the active site pocket. True or false?

14 An allosteric inhibitor closely resembles the substrate in shape, and inhibits reaction by combining with the enzyme at the active site. True or false?

15 Amino acids and sugars are transported across the intestinal cell membrane by combining with *permease* proteins in the membrane. Would you expect the same permease to operate for the transport of both amino acids and sugars?

16 Bacterial ribosomes are made up of two subunits, 50S and 30S, and each of these is composed of both protein and RNA. True or false?

17 Complete the sentences (i) to (iv) below by inserting the most appropriate of the following terms: DNA, mRNA, tRNA, polypeptide.
(i) ... is the principal nucleic acid constituent of chromosomes.
(ii) Genetic information passes from nucleus to cytoplasm in the form of ...
(iii) Amino acids are held in place on the ribosome, next to the growing polypeptide chain, by means of their specific ...
(iv) In the Jacob–Monod hypothesis, synthesis of β-galactosidase protein is inhibited when repressor protein binds to ...

18 Complete the following sentences by inserting the words 'parallel' or 'perpendicular' as appropriate.
(i) The hydrogen bonds in an α-helix run ... to the main polypeptide chain.
(ii) The hydrogen bonds in a β-pleated sheet (e.g. silk) run ... to the main polypeptide chain.

Study guide for Unit 1

The most important part of this Unit is Part I, which contains material essential for the rest of the Block and indeed for the whole Course. None of this can easily be omitted, but if pressed for time you may cut down on Part II, e.g. the short Section on light scattering (Section 1.6) or the long Case Study (Section 1.8), if really necessary. However, the concept of sedimentation coefficient (Section 1.7.2) should not be left out as it comes up again many times in this Block.

The model building kit is helpful though not essential for Sections 1.1.1 and 1.3.2, and the first television programme provides substantial backing for the whole of Part I.

You will notice a lengthy pre-Unit test at the beginning of this Unit. This should be studied carefully as it covers prerequisite material for this and subsequent Units of the Block. Anything in the pre-Unit test on which you need further information should be traceable from the references in Table A1.

For convenience, many items from previous Courses have been repeated in the *Source Book*.* There are frequent references (marked with a dagger, †) to this *Source Book* throughout the Unit.

Associated with the TV programme linked to this Unit is a set paper (Perutz, 1976**). This provides a general introduction to molecular biology and should be read before you view the programme.

* The Open University (1977) S322 *Biochemistry and Molecular Biology: Source Book*, The Open University Press.

** M. F. Perutz (1976), Life with living molecules, *New Scientist*, **72**, 144–147.

Part I Introducing macromolecules

1.1 Properties of macromolecules

Study comment This Section begins with a definition of macromolecules designed to encompass all four classes of biochemicals, and then goes on to point out the differences between them—e.g. is their role informational, or do they play a more passive part as structural or storage molecules? From then on, the emphasis is on similarities rather than differences, beginning with a recapitulation of structural hierarchy (primary to quaternary structures), followed by a consideration of those properties which may be said to be characteristic of macromolecules in general— specificity, flexibility and fragility. Many of the examples are taken from proteins or nucleic acids rather than polysaccharides or lipids. This is partly deliberate, as you will already have covered their structure in sufficient detail in an earlier Course[1]*; you should refer back to this Course for revision if necessary. Polysaccharides and lipids will tie in with the present Unit much better after you have read Unit 3, but at this stage you should decipher enough of the relevant formulae in the *Source Book* to see at least how they fit into the general structural hierarchy.

1.1.1 What is a macromolecule?

You are already familiar with the idea that a macromolecule is formed by polymerization of a large number of *building blocks*** like amino acid or glucose residues. The elements of water are removed from two such combining residues to give a covalently bonded structure, as shown in Figure 1. As more and more residues are added, the final molecular weight may run into several millions. This format holds for proteins, nucleic acids and polysaccharides but has to be somewhat slanted to accommodate the lipids, where the building blocks are very much larger. Theoretically, these lipid building blocks are derived from bridging molecules like glycerol, by condensations between fatty acid or phosphoric acid derivatives and the three hydroxyls on the three 'arms' of the glycerol E (see Figure 2). This substituted glycerol may be quite a large molecule— tristearin‡, for example, has a molecular weight of 891 but it cannot be denied that it is still small compared with, say, a protein. In biology, however, lipid molecules like this are rarely if ever found as individual entities, but are clumped together in specific molecular aggregates—macromolecules, in fact. What holds the aggregates together is non-covalent bonding, as we shall describe in more detail in Unit 3, and it is this that distinguishes lipids from other macromolecules. In proteins, nucleic acids and polysaccharides, of course, the building blocks are held together by *covalent* bonds. (Return to this description after Unit 3, and see if you still agree that lipids should be classed as macromolecules.)

polymerization

building blocks

NOW YOU COULD TRY SAQS 1 AND 2 ON P. 48.

You should now study the formulae of the various building blocks in the Source Book (under amino acid, nucleotide, monosaccharide and lipid). Many of them have been described in previous Courses, and the main point to appreciate here is the large number of different building blocks found in each class and therefore the correspondingly enormous number of variations which are available through polymerization or aggregation.

At this stage it would be instructive to use the model building kit to build a small dipeptide (e.g. alanylglycine) and a disaccharide. This should impress on you the relative sizes of the macromolecular building blocks, and also their great variety.

* Superscript numbers are references to other Open University Courses which are listed on p. 47.

** Many texts, including S2–1, use the term *monomer* instead of building block. Unfortunately, 'monomer' is also used in quite a different context, to describe the components of a multimeric protein. We have therefore tried to steer clear of it altogether, and to use the more naive term 'building block'.

‡ Tristearin is a triglyceride†, where R is $-CH_3(CH_2)_{16}$.

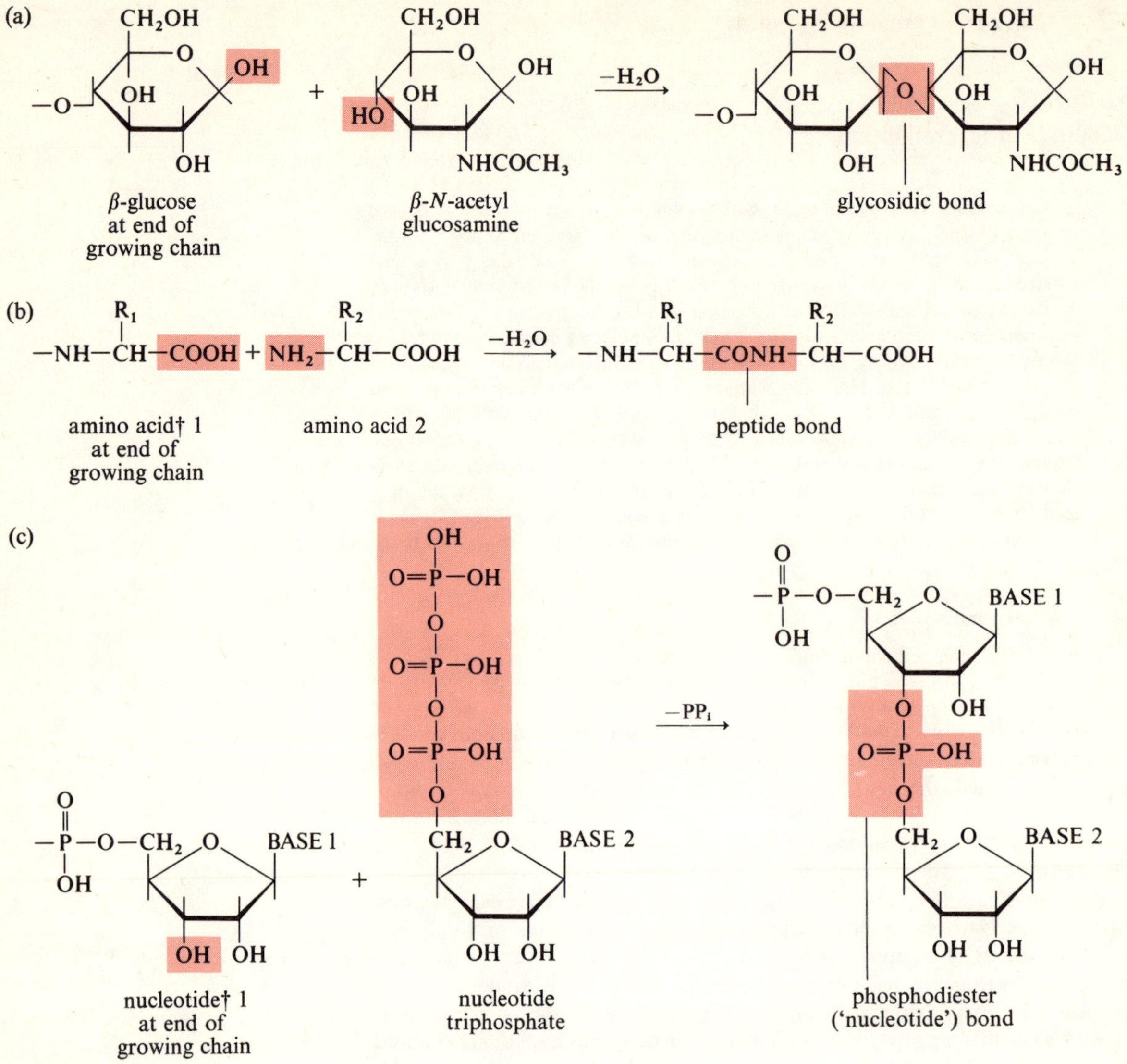

(a)

β-glucose
at end of
growing chain

β-*N*-acetyl
glucosamine

glycosidic bond

(b)

amino acid† 1
at end of
growing chain

amino acid 2

peptide bond

(c)

nucleotide† 1
at end of
growing chain

nucleotide
triphosphate

phosphodiester
('nucleotide') bond

Figure 1 Polymerization of macromolecular building blocks. (a) Polysaccharides, (b) proteins, (c) nucleic acids. The parts of the molecule involved in bond formation are shaded in red. (The significance of the term β in polysaccharides will become apparent in Unit 3. For proteins see further details on R_1 and R_2 under amino acid in the *Source Book*, and for nucleic acids see the *Source Book* under bases. Note that RNA not DNA is shown here. The difference between the two nucleic acids, and the full formula of pyrophosphate (PP_i) can be found in the *Source Book* under nucleic acid.)

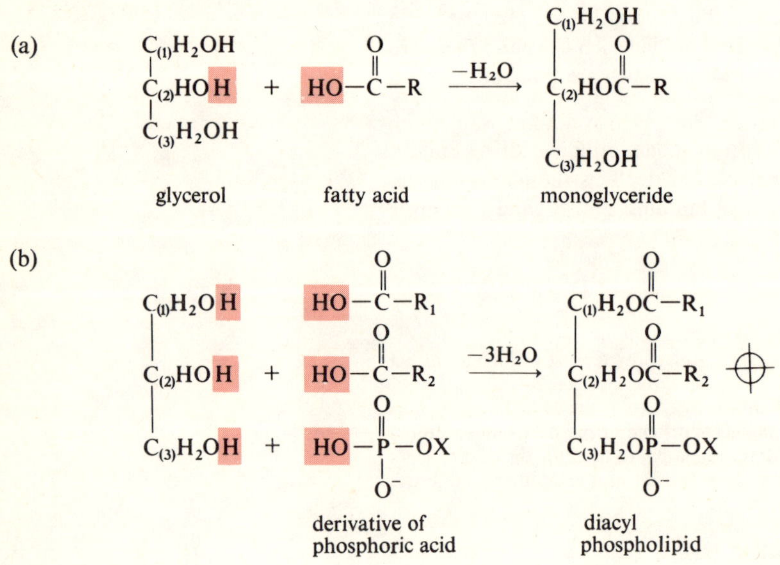

(a)

glycerol

fatty acid

monoglyceride

(b)

derivative of
phosphoric acid

diacyl
phospholipid

Figure 2 Lipid building blocks. (a) Formation of a neutral fat, (b) formation of a phospholipid. (R, R_1 and R_2 are long fatty acid chains, see under fatty acid in the *Source Book*. For the nature of X groups see under phospholipid in the *Source Book*. $C_{(1)}$ and C-1 are alternative ways of labelling the first carbon atom in a chain.)

1.1.2 Informational, structural and storage roles of macromolecules

Proteins and nucleic acids are often called informational molecules, in reference to their job of conveying information by means of highly specific interaction with other molecules. In both cases this information is carried in the sequence of building blocks that go to make up the primary structure, but it is only with nucleic acids that the primary structure can be seen to have a *direct* role to play. Interaction between mRNA and tRNA during translation† of the genetic code is a good example. With proteins the primary structure acts only indirectly, by dictating the pathway of protein folding to give the unique 3-D shape of the native* molecule. This shape fulfils the informational role by providing binding sites tailor-made for the other molecules with which the protein interacts. Some examples of this are given in ITQ 1.

informational macromolecule

> **ITQ 1** In which of the proteins in the right-hand column of the Table would you expect to find a binding site specific for the molecules in the left-hand column? (Answer on p. 52).

Molecule binding to protein	Protein
(a) substrate†	enzyme†
(b) antigen	carrier protein or permease†
(c) competitive inhibitor†	antibody
(d) 'passenger' molecule†	receptor† protein
(e) allosteric† inhibitor	
(f) neurotransmitter†	
(g) hormone†	

> *Note* The terms 'antigen' and 'antibody' will be more precisely defined in Block F. For the present, we may say simply that an *antigen* is any foreign particle which is capable of stimulating the release of specific neutralizing antibodies when introduced into the body of a higher animal, while the *antibody* is a protein with a binding site designed specifically to entrap that particular antigen and effectively remove it from circulation.

As you can see from **ITQ 1**, and later in Figure 5, some proteins may have specific binding sites for more than one type of molecule; for example, a membrane protein may well bind to lipid (as part of its structural role) and also to hormone, passenger molecule or neurotransmitter, depending on its biological function. The important point to remember is that many of the highly specific interactions characteristic of living organisms can be tracked down, at the molecular level, to a specific binding site on some protein. You should find many examples of this throughout the Course, and the whole subject of receptors is discussed in Unit 12.

binding sites

Although globular proteins are undoubtedly informational molecules, the fibrous proteins described in an earlier Course[1] are rather different, resembling in some ways the polysaccharides and lipids, which are mainly storage or structural molecules. Storage molecules have much less need for specificity than informational molecules, and there is therefore no point in indulging in large numbers of different building block types when mere repetition of the same formula will do just as well. The add on/take off structure of glycogen[2] is a good example of this, and also demonstrates that there is no particular need for all molecules of a particular type of storage compound to have the same size. The resulting heterogeneity of molecules in the population is one thing which has until recently deterred research workers from structural studies on polysaccharides.

structural/storage macromolecules

heterogeneity

NOW YOU COULD TRY SAQ 3.

---

* Native is defined in the *Source Book*. Briefly, it means the biologically active conformation of a macromolecule.

1.1.3 Structural hierarchy

The following brief description of primary, secondary and tertiary structure is intended to underline what you have already covered in an earlier Course[3] rather than to break new ground. Detailed descriptions of secondary and tertiary structure of both proteins and nucleic acids appear in Unit 2, but meanwhile Figure 3 is a good summary of the information required for this Unit. The account of quaternary structure is expanded to define a number of terms that will reappear throughout this Unit.

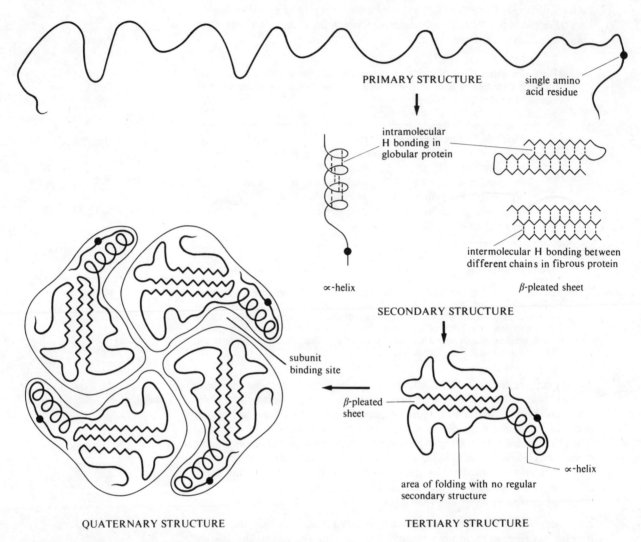

Figure 3 Structural hierarchy in a protein macromolecule.

Primary structure describes the order of covalently bound building blocks* in the backbone, while *secondary structure* describes the way in which certain lengths of backbone fold into recognizable hydrogen bonded** elements such as the α-helix and the β-pleated sheet in proteins, and the double helix in nucleic acids. *Tertiary structure* describes how these elements, together with certain other less well-defined folding patterns, are incorporated into the final 3-D shape (see Figure 3).

primary structure
secondary structure

tertiary structure

---

* In this Block we are concerned mainly with 'pure' macromolecules where all the building blocks are of one type—all amino acid or all monosaccharide for example. However, there are plenty of covalently linked *hybrid* macromolecules in which short stretches of a different macromolecular type are covalently bound to the main backbone. Common examples are the glycoproteins (protein plus polysaccharide) and the lipopolysaccharides (polysaccharide plus lipid). These should not be confused with the non-covalently linked multimeric complexes described later in this Unit.

** See Figure 6b. Hydrogen bonding is defined in Section 1.2.2.

Up to this point we have assumed that we are talking about a single molecule with a covalently bonded backbone. However, a complete macromolecule may consist of several such units all specifically interlocking, and the way in which they do so is known as the *quaternary structure*.

quaternary structure

Each interlocking unit with its single covalently bonded backbone is known as a *subunit*. In proteins, for example, the subunit is a single polypeptide chain. There may be comparatively few of these as in haemoglobin, which has two types of subunit, α and β, arranged to give quaternary structure $\alpha_2\beta_2$. On the other hand, viruses and ribosomes (better known as self-assembling systems) may have hundreds of interlocking units all linked together on the same specific interaction principles as haemoglobin. Membranes are built up from the same principles again, but here it is difficult to know quite where one molecule ends and the next begins.

subunit

QUESTION A simple virus, as shown in Figure 4d, may be composed of a single molecule of nucleic acid surrounded by a large number of identical protein molecules. This structure has two different types of (i) building block and (ii) subunit. Name these.

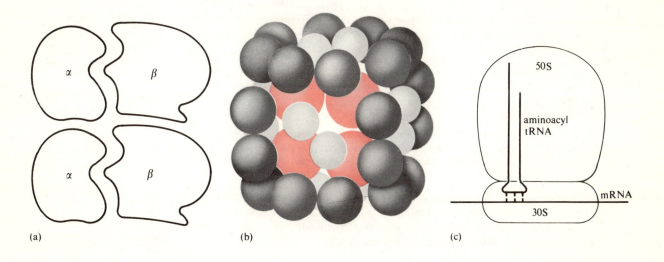

(a)

(b)

(c)

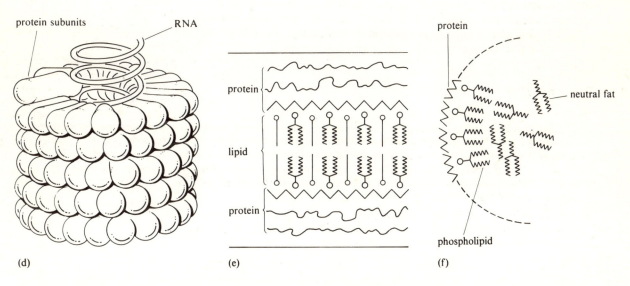

(d)

(e)

(f)

Figure 4 Multimeric complexes. (a) Haemoglobin, with four protein subunits. (b) Tentative model of pyruvate dehydrogenase complex (each ball represents a group of protein subunits forming one active enzyme). (c) Ribosome (both 50S and 30S 'subunits' are composed of non-covalently linked protein and RNA; there is also non-covalent bonding between mRNA and tRNA at the site of protein synthesis†). (d) Tobacco mosaic virus, with 2 100 protein subunits and one RNA subunit. (e) Tentative model of cell membrane. (f) Tentative model of a chylomicron (the lipoprotein particles in which digested fat is transported in the blood stream).

ANSWER (i) *Amino acids* are the building blocks of the protein components, and *nucleotides* of the nucleic acid component.

(ii) The *nucleic acid molecule* is one type of subunit—a single polynucleotide chain formed of covalently linked nucleotide building blocks. This type of subunit occurs only once in the whole virus, but the other type of subunit, the *protein molecule*, is repeated many times.

QUESTION Complete the following sentence: The protein haemoglobin has a total of ... subunits of ... different types.

ANSWER There are *four* subunits (two α, plus two β) of *two* different types (α and β). The important point to remember is that a subunit is a single polypeptide (or polynucleotide, polysaccharide, etc.) chain.

A term you may come across in descriptions of these giant complexes is *multimeric complex*. This may be defined as a collection of molecules held together at specific interaction sites (subunit binding sites) by non-covalent bonding. The constituent molecules or subunits are often difficult to dissociate from each other and once dissociated may no longer have any biological activity. Large multimeric complexes, like pyruvate dehydrogenase complex (PDC) shown in Figure 4b, may dissociate incompletely, leaving groups of subunits that still retain their enzymic activity. In PDC these active groups are of three different types, corresponding to the three constituent enzymes of the complex. We shall go into this in more detail in Section 1.8.

Subunits are by no means always composed of protein. Nucleic acids, lipids and polysaccharides can equally well form part of a multimeric complex, as you can see in Figure 4.

ITQ 2 Using the information in Figure 4, state what classes of macromolecule form the subunits of the following multimeric complexes: (a) the 50S ribosomal 'subunit'; (b) a chylomicron; (c) a virus; (d) a cell membrane.

An important point to remember is that of the entire structural hierarchy, only the primary structure is held together by covalent bonds. The 3-D shape (secondary and tertiary structure) and the specific interactions with other subunits (quaternary structure) are all dependent on weak bonding.

Also dependent on weak bonding are the interactions shown in Figure 5, i.e. those between macromolecules and other molecules which fit into their specific binding sites—the kind of interaction we mentioned in ITQ 1. Therefore, not only is the secondary to quarternary structure of each macromolecule dependent on weak bonding, but so also is its interaction with other molecules, large and small. It is now time to focus on these bonds in more detail.

1.2 Weak bonds

Study comment We have assumed that you are already familiar with H-bonded structures in protein and nucleic acids (see pre-Unit test question 18; you should revise the relevant Section of an earlier Course[1] if necessary). We hope that you will return to this Section after reading about polysaccharides and lipids in Unit 3. By then, you should be able to give specific examples of the different types of weak bond from all classes of macromolecules.

The main aim of this Section is to show how non-covalent bonding accounts for many of the unique properties of macromolecules (Objective 4) and to appreciate this you should refer throughout to the data in Table 1.

This Section begins by emphasizing the difference between covalent and non-covalent bonds in terms of bond energy and stability. It then describes in more detail the hydrogen, van der Waals and ionic bonds, and points out where the hydrophobic bond fits in. Weak bonds are particularly susceptible to their environment, and ionic bonding has been chosen to give specific examples of this.

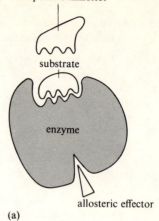

(a)

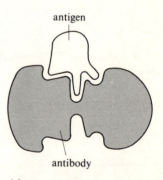

(b)

(c)

Figure 5 Specific interaction between macromolecules and small molecules. (a) Enzyme complexes with substrate†, competitive inhibitor† and allosteric† effector; (b) receptor protein complexes with hormone and neurotransmitter†; (c) antigen–antibody complex (note that antigens may be large or small molecules, see Block F).

1.2.1 Covalent versus non-covalent bonding

What are these weak bonds that have such profound effects on the properties of macromolecules? The three most relevant to biology are the hydrogen, van der Waals and ionic bonds. All of these you have met before, but there are two reasons for re-introducing them here: to give specific examples for all four classes of macromolecules, and to put things on a more quantitative basis. With covalent bonds, the electron clouds† around each of the combining nuclei fuse to create a new electron cloud, the *molecular orbital*. The bond energy of such an arrangement is high, around 200–400 kJ mol⁻¹. Hence covalently bonded structures tend to be rather stable, and the breaking of such bonds requires either extremes of temperature and pH or, as in cellular metabolism, enzymic catalysis. With non-covalent bonds the picture is quite different. There is no merging of atomic orbitals, but rather an electrostatic attraction between positive and negative areas of combining atoms. The result is a much weaker bond which has the enormous biological advantage that it can be made or broken without any major energy change. If you realize that the kinetic energy of molecules in solution is around 2.5 kJ mol⁻¹, and then look at the bond energies in Table 1, you will see what we mean.

Bond energy is defined as the energy required to break the bond, and not unexpectedly it is related to the distance between combining atoms. The relationship is:

$$F \propto \frac{1}{\epsilon d^n} \qquad (1.2)$$

where F is the force of interaction between combining atoms, d the distance between them, and ϵ (epsilon)* the dielectric constant† of the medium. The symbol n is a constant for a given bond type (see Table 1**); this is another way of saying that for some bonds the combining atoms need to be closer together than for others, before there is any effective force between them.

margin note: covalent bonding

margin note: weak bonding

margin note: bond energy

TABLE 1 Characteristics of bond types found in macromolecules

Bond type	Bond energy, F /kJ* mol⁻¹	Bond length, d /nm	Value of n**
covalent	200–400	0.10–0.15‡	—
hydrogen	up to 20	0.26–0.31	4
van der Waals	up to 4	0.40–1.00	7
hydrophobic	4–8	0.40–1.00	—
ionic	up to 4	—	2

* 1 kcal = 4.18 kJ.
** Where $F \propto 1/\epsilon d^n$.
‡ For organic molecules.

1.2.2 Hydrogen bonds

In biology, hydrogen bonds may be found wherever there is an opportunity for the hydrogen proton to be shared between two electronegative† atoms, usually oxygen and nitrogen. Directionality is also important, and the bond energy is greatest if it can be arranged that the three combining atoms lie in a straight line (see Figure 6a‡). This is one reason why the α-helix is such a favoured form of

margin note: directionality

---

* A list of Greek letters with their English pronounciation is given in the Table of Constants and Symbols in the *Source Book*.

** In column 3 you will see the expression 'Bond length/nm'. This means that bond length (which was measured in nm) is *divided* by nm, so that the figures in the column are pure dimensionless numbers. *In other words, we are using 'Bond length/nm' instead of the old form, 'Bond length (nm)'.* You will find this system employed in many recent scientific books and papers, and in future OU Science Courses.

‡ This can also be seen in Stereoslides 4 and 5. Slide 4 shows how a single amino acid residue (outlined in orange) fits into a complete helix, and you can also see the H bonds (in red) running parallel to the helix axis. Slide 5 shows a section of helix where the H bonds have been taken out. Note how unstable the structure appears. (Slide 5 is constructed from an older model building kit than Slide 4; oxygen is red, and nitrogen blue.)

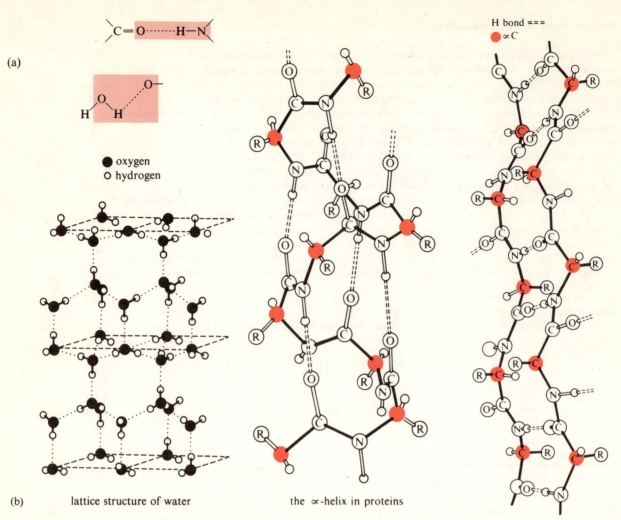

(a)

• oxygen
○ hydrogen

(b)

lattice structure of water   the ∝-helix in proteins   H bond ===
● ∝C

Figure 6 Hydrogen bonding in macromolecules. (a) Importance of directionality, (b) H bonding in biological structures (dotted lines represent H bonds).

the β-pleated sheet in proteins (silk)

secondary structure. Every turn of the helix takes up 3.6 amino acid residues (i.e. 3.6 units of $-NH-CHR-C-$) and this brings the **NH** of the peptide bond

exactly in line with the **C=O** below. Look at the structure of silk shown in Figure 6b, to convince yourself that this linear arrangement of $O-H-N$ is true also of the β-pleated sheet.

H bonding also stabilizes secondary structure in polysaccharides, as you will see in Unit 3. (In the *Source Book*, you will note that there are numerous sidechain OHs protruding from the monosaccharide rings.) In nucleic acids it is the hydrogen bonds that are responsible for specificity of interaction—as in replication† and translation†—while most of the binding energy comes from hydrophobic bonding (see Unit 2). In lipids the H bonds are somewhat overshadowed by hydrophobic bonding again, but both hydrogen and ionic bonds are prominent at the polar head end of the molecule. In proteins, we have just described how H bonding from the elements of the peptide bond stabilizes secondary structure, but H bonding from amino acid sidechains is also important in stabilizing tertiary structure.

ITQ 3 Without looking at Figure 7, select those residues available for H bonding in each of the following molecules (a)–(d). (Formulae can be found in the *Source Book* under monosaccharide, amino acid, lipid or nucleotide.)

(a) Glucose, galactosamine, galactose-6-sulphate;
(b) Dipeptides formed from lysine+leucine, and from alanine+tyrosine;
(c) Cytosine, guanine;
(d) Stearic acid.

Hint Any H attached to N or O is a likely candidate, provided there is space for an incoming N or O to complete the H bond. Similarly, any N or O with space for an incoming H—O or H—N will do.

18

1.2.3 van der Waals and hydrophobic bonds

The van der Waals bonds we shall be concerned with here are those formed between so-called *non-polar*† residues, i.e. those molecules or parts of molecules where the electron clouds are symmetrically distributed and there are no local charge differences.

non-polar residues

> **ITQ 4**   Classify the following residues into polar and non-polar:
>
> $$-CH_3, -CH_2-, -C=O, -O-H$$
>
> *Hint*   Bonding between atoms of unequal electronegativity is what produces a polar molecule.

Although the isolated non-polar molecule is electrically neutral, local charge differences can be *induced* in the following way. The electron cloud may become transiently asymmetrical, creating small local inequalities of charge or *dipoles*. If such a transient dipole is now approached by another group also transiently polarized but in the reverse direction, the result is a net attractive force between them. It is this force which is known as a *van der Waals bond*.* As you can see from Table 1, the bond energy falls off rapidly if the combining atoms draw apart by even a small amount. This has important consequences for specificity of inter-action. Molecules held together largely by van der Waals bonds have to approach extremely closely before there are enough contacts to give a significant net binding force between them. Any small protuberances or mismatches will seriously reduce the number of possible van der Waals contacts and the bond energy will fall drastically. This is one reason why macromolecules may fail to 'recognize' any molecule that does not have the exactly complementary shape. The extreme specificity of antigen–antibody interaction, for example, relies largely on such a mechanism.

dipoles

complementary shape

One of the most important bond types to be found in biology is the *hydrophobic bond*, which derives part of its energy from van der Waals bonding. Hydrophobic interactions occur wherever non-polar residues coalesce, squeezing out the water molecules trapped between them. This permits two types of bond to be formed simultaneously—van der Waals bonds arise between the non-polar groups left in the molecule interior, and hydrogen bonds form as the H_2O molecules released from the interior go to extend the 3-D lattice of water molecules in the surrounding medium. Figure 6b shows such a lattice. Ordering of water molecules in the cell interior is one of the topics to be discussed in Unit 7; at this stage you should merely note it as a major contributor to the net binding energy of hydrophobic bonds.

Hydrophobic bonds are formed between non-polar residues in proteins (see Ala, Trp, etc., under amino acids in the *Source Book*) and between the stacked purine and pyrimidine rings of nucleotide† bases in nucleic acids. Similar bonding possibilities exist in the polysaccharides, but it is amongst the lipids that hydrophobic bonding really comes into its own (see Unit 3).

> **ITQ 5**   Without looking at Figure 7, outline the non-polar residues in the macromolecules listed in ITQ 3.

1.2.4 Ionic bonds

These tend to be found on the surface of macromolecules rather than in the hydrophobic interior, and they can operate over much greater distances than van der Waals bonds.

> **ITQ 6**   What piece of information in Table 1 can be used in confirmation of this last statement?

---

* Strictly, what we have just described is a *London dispersion force*, which may be defined as an attraction between two temporary dipoles. A van der Waals bond may also receive contributions from permanent dipoles, but these are of only minor importance in the biological situations we are to describe.

LIVERPOOL INSTITUTE OF
HIGHER EDUCATION
THE MARKLAND LIBRARY

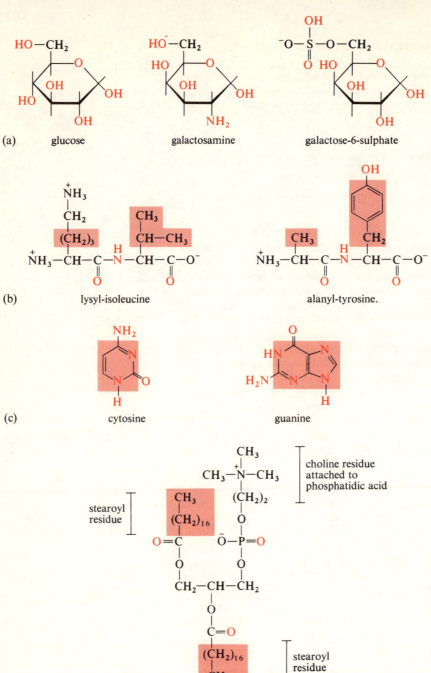

Figure 7 Weak bonding capabilities of specific groups in macromolecular building blocks. (a) Monosaccharides, (b) dipeptides, (c) nucleotide bases, (d) phospholipid. Potential H-bonding residues are printed in solid red, and potential van der Waals bonding residues as red shaded areas. However, the red shading has been omitted from some —CH₂— groups where nearby polar residues would impair van der Waals interaction.

There are fewer opportunities for ionic bonding in macromolecules than for H or hydrophobic bonding, because only fully ionized groups can participate. These groups are not always easy to spot from the formulae of macromolecular sidechains, particularly as they may be written in their un-ionized forms.

ionizable groups

ITQ 7 Write down the reactions responsible for producing ionized groups from the following residues:

(a) carboxyl $-COOH$ (b) sulphate (c) phosphate (d) amino $-NH_2$ (e) histidyl (an amino acid residue)

Hint Look at the dissociation reaction of an acid, described in the *Source Book* under pK.

At biological pH values, both phosphate and sulphate groups are nearly always fully ionized. Carboxyl and sulphate groups are prominent in structural poly-saccharides like hyaluronic acid (which lubricates knee joints), while phosphate groups play a strong part in the orientation of phospholipids in membranes.

The molecular environment or *microenvironment* is important in ionic bonding, particularly if the groups concerned are only partially ionized at biological pH values. This is because the degree of ionization of the two interacting groups has a profound effect on binding energy, for reasons which will become clearer if we rewrite equation 1.2 in the form that applies specifically to ionic bonding:

microenvironment

$$F = \frac{Q_1 Q_2}{d^2} \qquad (1.3)$$

Here, Q_1 and Q_2 are the electrostatic charges on the two combining atoms and it is these that are so very dependent upon the degree of ionization. This in turn depends on the environment. With an enzyme protein, for example, not only structure but also activity may be influenced by this environment. If the ionizable group lies in the active site and is directly responsible for the catalytic event, its degree of ionization will obviously affect the overall enzyme activity; this partly explains the bell-shaped pH–activity curve you may have encountered in earlier Courses. In just the same way, pH can affect the binding of substrate, hormone, tRNA, antigen, or whatever binding activity is the business of that particular macromolecule.

effect of pH

NOW YOU COULD TRY SAQS 4, 5 AND 6.

1.3 Specificity, flexibility and fragility of macromolecules

Study comment This Section relates to Objective 4 and describes three charac-teristic properties of macromolecules—specificity, flexibility and fragility. You should throughout be looking for examples to show how these properties depend upon the presence of weak rather than covalent bonds.

The Section on specificity begins with an example, and then attempts to explain specificity in terms of weak bonding. Flexibility is described in terms of conforma-tional changes. The molecular basis for this is given in Figure 8, and several examples of biochemical control mechanisms involving such conformational changes then follow. Fragility is described in terms of denaturation which in turn leads into a discussion of protein folding and unfolding—a topic that will be ex-panded in Unit 2.

1.3.1 Specificity

Specificity is a characteristic of all macromolecular reactions, including reactions with other macromolecules (as in membranes, mitochondria,[A]* multi-enzyme complexes, etc.) and with small molecules (like hormones, substrates, neuro-transmitters, and 'passenger' molecules for transport). In all cases specificity comes about simply because macromolecules recognize targets by their shape and their chemical reactivity. This is thought to be one reason why the vast majority of naturally-occurring amino acids occur as the L optical isomer rather than the D form.

recognition by shape

MAKE YOURSELF A MODEL OF D- AND L-LACTIC ACID† (EVEN THOUGH THESE ARE SUGAR DERIVATIVES, NOT AMINO ACIDS OF COURSE), AND COMPARE THEIR SHAPES.

This should convince you that to an enzyme binding site, for example, these are really very different molecules. It is then not difficult to accept that any enzyme capable of recognizing, say, L-amino acids, will be at least partially blind to D-amino acids. This is an extreme example of specificity, but it may explain why, once started on the L-amino acids, life seems to have selected for this rather than the D form. This does not, of course, explain how the initial choice was made, but certainly it is true that D-amino acids are rare except in bacteria.

---

* Reference for items marked with a superscript A are to be found in Table A1, which should be consulted if you require further information.

How does weak bonding come into specificity? For specific interaction, the tertiary structure of the macromolecule must include a binding site of shape complementary to that of its interacting partner. Specificity of interaction lies in the fact that *a very large number* of non-covalent bonds are needed to build up a binding force of any size. This in turn necessitates a very close fit, such as we have already described for van der Waals bonds. These two interlinked factors—closeness of fit, and the large number of weak bonds necessary for effective binding—are key concepts in specificity.

NOW YOU COULD TRY SAQ 7.

closeness of fit

1.3.2 Flexibility

All elements of the structural hierarchy that rely on weak rather than covalent bonding can readily be broken and remade. This fact has important consequences for both the fragility and flexibility of macromolecules. *Flexibility* may be thought of as the ability to undergo *conformational changes*, and these in turn may be defined as small changes in shape brought about with only minimal changes in energy.

conformational change

The molecular basis of these changes is as follows: Consider the hypothetical small molecule M, shown in Figure 8. Both C-1 and C-2 are carbon atoms, while Y is any other atom with at least one spare valency, such as $-\overset{|}{N}-$, or $-O-$. The angle between the three covalently bonded atoms C-1, C-2 and Y is rigidly defined, with very little scope for overlap or compression; the only degree of flexibility comes by rotating Y about the single bond[A] between C-1 and C-2.

(a) (b)

Figure 8 Role of weak bonds in conformational changes of macromolecules. C-1 and C-2 are carbon atoms, Y is any divalent atom, and X is a bulky molecule or part of a molecule which is bound to Y by (a) covalent bonding, or (b) non-covalent bonding.

It is this single bond rotation that is exploited in conformational changes of macromolecules. Imagine that Y is covalently bound to another very bulky group, X (Figure 8a). No free rotation would then be possible. However, if Y is only *weakly* bound to X (Figure 8b), very little energy is needed to break the X—Y bond, leaving Y free to rotate about the C-1—C-2 bond.* It is this kind of mechanism at the molecular level that is the basis of conformational changes in macromolecules.

Some instances of conformational changes may already be familiar to you. The active site† of an enzyme, for example, may alter shape on interaction with the substrate (the induced fit† theory of enzymic catalysis), and this brings catalytic groups into closer contact with vulnerable bonds in the substrate. This is illustrated in Figure 9a, which shows a particularly dramatic example such as you might meet in carboxypeptidase, one of the digestive enzymes.** Here, binding of dipeptide (its smallest substrate) causes tyrosine (a key residue on the surface of the molecule) to swing in from its position on the surface of the molecule, moving through 120° to end up close to the peptide bond of the substrate. Such flexibility could hardly occur if the molecule were totally constrained by covalent bonds. Figure 9b shows an allosteric† protein, another case where flexibility is

---

* A model of a molecule like M (Figure 8), in which Y is an oxygen atom, may be useful here. Try using the model building kit to make

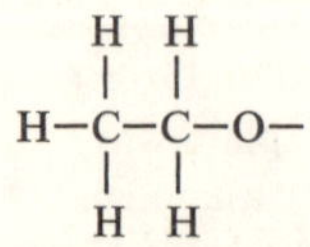

$$\begin{array}{c} H \quad H \\ | \quad | \\ H-C-C-O- \\ | \quad | \\ H \quad H \end{array}$$

and imagine the rotation about the C—C bond when the —O— has (a) a strong bond and (b) a weak bond, to another very large molecule.

** Stereoslides 1 and 2, to be described in Unit 2, also illustrate this point. From a distance, as in this view, you can see that the tyrosine (purple residue prominent on top of the molecule) in Slide 1 has disappeared into the interior of the molecule in Slide 2.

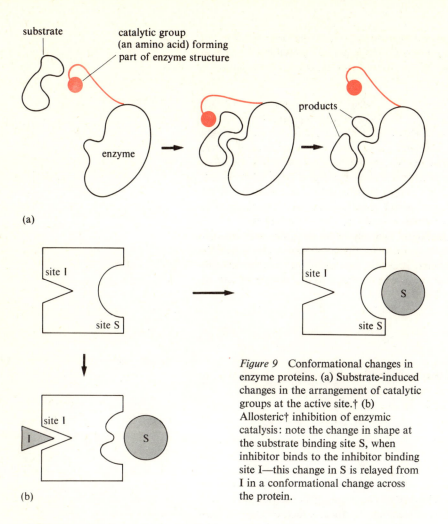

(a)

(b)

Figure 9 Conformational changes in enzyme proteins. (a) Substrate-induced changes in the arrangement of catalytic groups at the active site.† (b) Allosteric† inhibition of enzymic catalysis: note the change in shape at the substrate binding site S, when inhibitor binds to the inhibitor binding site I—this change in S is relayed from I in a conformational change across the protein.

all important. Here, events at one binding site can influence those at other, often remote, sites on the molecule. The precise series of molecular events responsible for relaying this information is so far known only for haemoglobin (probably the world's most intensively studied allosteric protein) but in all cases transmission of information across the molecule must take place by the making and breaking of weak bonds.

Proteins are by no means the only flexible macromolecules. Both DNA and RNA can exist in more than one conformation, as you will learn in Unit 2, but the relationship between conformational change and biological role is as yet less clear-cut than in proteins. In polysaccharide chemistry, control through conformational change is just becoming accepted; this is a new and exciting area we shall mention very briefly in Unit 3.

One further example of control through conformational change is the built-in *obsolescence* found in many macromolecules. Certain bonds in the primary structure are particularly vulnerable to degradative enzymes of the cell. Normally these are 'hidden' by folding of the tertiary structure or by interaction with other molecules, but conformational changes under controlled conditions may expose them to attack. Again this idea is relatively new, and there are few concrete examples. But the point to remember is that here, as in all conformational changes, weak bonding in all types of macromolecule permits a flexibility not found in other structures. This in turn opens up a method of controlling reactions which is unique to biological systems.

obsolescence

1.3.3 Fragility

Finally, we come to the third characteristic property of macromolecules—*fragility*. This phenomenon is well known to chemists working with macromolecules, especially when trying to purify a compound without destroying its native conformation. The thing to avoid is *denaturation*. This may be defined as partial or complete unfolding of the macromolecule backbone due to loss of weak-bonded elements of structure, i.e. secondary, tertiary and quaternary structure.

denaturation

Denaturation can usually be detected by the accompanying loss of biological activity, but a more direct approach is to follow the unfolding itself by any method sensitive to conformation. Spectroscopy (Unit 2) or viscometry (Section 1.5.1) are good examples.

It is not surprising that the army of non-covalent bonds holding the molecule together should have a tendency to fall apart once the very precise conditions of the cell environment are disrupted.* However, denaturation is no longer thought of as an all-or-nothing process. Intermediate states can be demonstrated, where partially unfolded protein is relatively stable and may even retain some of its biological activity. Perhaps this is not surprising. Denaturation is after all only an extreme case—a conformational change that went too far. Since the higher elements of structure in proteins are entirely determined by primary structure in a given environment, it is in principle perfectly possible to refold even a fully denatured (*random coil*) macromolecule. One need only simulate the exact conditions of the cell environment, since these are the conditions under which the molecule folded up in the first place. It should then spontaneously refold to its native conformation. This situation is obviously not a simple one to reproduce, and is discussed again in Units 2 and 4.

NOW YOU COULD TRY SAQ 8.

Part II Size and shape of macromolecules

Introduction

Up to now, we have described the properties of macromolecules without giving any idea of how this information was obtained. How do you know that what we say is true? To give even the smallest hope of answering this question we need to digress for some time into details of technique. Although methodology in itself may not seem very interesting, much of the mystique of modern science could be cleared away if only outsiders could understand something of what was going on. For example, what happens behind the impassive exteriors of ultracentrifuges, spectrophotometers and other black boxes into which biochemical compounds are fed? Although it is obviously impossible to go into the finer points of the methods to be described, we hope that you will at least gain a feeling for the difficulty of applying the techniques of physics and chemistry to macromolecules, and for the sort of information that can be obtained. In this Section we shall introduce the two areas of macromolecule chemistry where we have chosen to emphasize technique, namely purification, and determination of size and shape.

Purification we shall only gloss over, with a brief discussion of one of the problems of dealing with macromolecules in solution, namely homogeneity. We shall then concentrate on two parameters** widely used for characterizing macromolecules—size and (to a lesser extent) shape. These are chosen not just for cataloguing purposes, but for the very good reason that neither the structure nor indeed the function of a molecule can be understood without some idea of its size. For those with a cataloguing bent, however, methods of determining molecular weight may be classified into direct and indirect. As with the chemistry of small molecules, most of the methods used are *indirect*, which means that it is not molecular weight itself that is measured but some physical property directly dependent upon it. Rates of migration under the influence of an electric field (electrophoresis†) or under a gravitational force (ultracentrifugation†) are common examples, and like all indirect methods they require the macromolecule to be in solution. This brings us immediately to the first problem—*non-ideality*.

indirect methods

* This is why purification of the enzyme hexokinase, described in television programmes of S2–1, concentrated on separation methods that avoided extremes of pH, temperature and ionic strength. In these programmes, you saw methods which tended to exploit small differences in charge and size by using chromatography, gel filtration and centrifugation under mild conditions simulating those within the cell.

** By 'parameter', we mean here any physical property capable of being measured (? metered).

24

In theory, an indirect method is no good unless the physical property being measured is related only to mass, i.e. molecular weight, and is not influenced by any other physical characteristic such as shape. Unfortunately, with the macromolecules there is no single measurement that fits these requirements. This is because the behaviour of macromolecules in solution is very much influenced by their sheer size; in thermodynamic terms, macromolecular solutions 'tend to be non-ideal'. This is why our description of indirect methods is preceded by a Section on *molecular excluded volume*, a key concept in non-ideality.

The problem of non-ideality tends to get worse the bigger the molecule, but sometimes size actually becomes an advantage. It is then possible to abandon indirect methods and view the molecule directly, using electron microscopy or X-ray diffraction. These *direct* methods of molecular weight determination are mentioned at the end of the Unit.

direct methods

The ground we are to cover is summarized in Table 2, and you should refer to this frequently in reading through the rest of the Unit. It should help to pinpoint just what type of information each technique can deliver—is it primarily to investigate shape, or can it be used to calculate molecular weight—and it also indicates whether a technique stands on its own or requires supplementary information in the form of D (diffusion coefficient), $[\eta]$ (intrinsic viscosity, eta), or standards of known molecular weight. Another point we have emphasized in the Table is just what is being measured experimentally. It is sometimes difficult to glean this information from more detailed descriptions of a method.

use of Table 2

We promised at the beginning of the Unit not to forget that for most biologists the techniques of chemistry and physics are only a means to an end. For this reason the description of technique is side-tracked in several places into a discussion of the structure of nucleic acids and their role in protein synthesis. This presupposes some knowledge from an earlier Course[5] which you should briefly revise if necessary. However, the main purpose of these interruptions is to point out the relevance of biophysical techniques to biology, and a more detailed discussion of protein synthesis is deferred until Unit 11. For similar reasons, the Case Study on pyruvate dehydrogenase complex also comes in an unexpected place. It was chosen to illustrate the use of two techniques—sedimentation velocity and electron microscopy—and is therefore placed between these two rather than at the end of the Unit.

1.4 Behaviour of macromolecules in solution

1.4.1 Homogeneity

For structural studies, a macromolecular preparation must be highly purified. This is no easy task, particularly when the macromolecule is to be identified at each stage by its biological activity. The purification techniques are then limited to those that will not denature the molecule by disrupting the weak bonds holding together its tertiary structure.

A further difficulty in purifying macromolecules concerns quaternary structure. In solution this may break down, producing a number of components from a single apparently pure macromolecule. Purification is then a question of knowing when to stop. Ideally, the answer is when the preparation is *homogeneous*, i.e. when all the molecules in it are identical. A homogeneous preparation should therefore show only one band when tested under conditions designed to show up any contaminating molecules, such as in ultracentrifugation or gel filtration where molecules separate according to size, or in electrophoresis where they separate according to charge.

homogeneity

Having said this, we come to two cases where a macromolecular preparation may be 'pure' but still not homogeneous. These are the structural/storage molecules (see Section 1.1.2) and the multimeric complexes (Section 1.1.3). A storage polysaccharide such as glycogen does not require a unique 3-D structure for its function, and a single pure preparation will probably contain a whole range of molecular sizes. An average value can be found, but the question of molecular weight is perhaps less relevant to polysaccharides than to proteins and nucleic acids. These need to be of a finite size for their informational role, so one would

TABLE 2* Techniques of measuring size and shape in macromolecules

Name	Type of macromolecule examined	What is measured experimentally	Outside information required	Information obtained	Space for notes
viscometry	1 protein, nucleic acid, multimeric complex 2 DNA	rate of flow of macromolecular solution relative to pure solvent	1 none 2 (a) standards of known mol. wt, or (b) sedimentation coefficient, s**	1 intrinsic viscosity $[\eta]$ giving information on *shape* 2 mol. wt	
sedimentation velocity	1 protein, nucleic acid, multimeric complex 2 protein 3 nucleic acid	rate of sedimentation	1 none 2 diffusion coefficient, D 3 intrinsic viscosity, $[\eta]$	1 sedimentation coefficient, s** 2 mol. wt 3 mol. wt	
sedimentation equilibrium	1 protein 2 nucleic acid in CsCl density gradient	1 distance travelled towards bottom of tube at equilibrium 2 density of caesium chloride solution in which particle comes to rest	1 none 2 none	1 mol. wt 2 buoyant density, ρ_B**	
light scattering	larger proteins, DNA	angle to which incident light is scattered by particle in solution, plus intensities of scattered light	none	mol. wt; radius of gyration R_G, giving information on shape	
gel filtration	protein	elution volume, V_e	standards of known mol. wt	mol. wt	
polyacrylamide gel electrophoresis	protein, RNA	electrophoretic mobility relative to marker dye	standards of known mol. wt	mol. wt	
microscopy	1 protein 2 DNA (usually with autoradiography)	size of image on photographic plate, and magnifying power of microscope	1 none 2 density per unit length	1 mol. wt; approximate shape; approximate number and arrangement of subunits 2 mol. wt; shape	

* This Table is a summary of what is to follow in the rest of the Unit. Therefore it may not be fully understandable at this stage. It is presented here so that you can begin to comprehend the relationships between the various methods and so that you will have the basic information in a convenient form for future reference.

** ρ_B and s are both used as they stand, to characterize a molecule and identify it for future reference. They are also used to detect different shapes of the same molecule.

expect that any pure preparation of protein or nucleic acid would indeed be homogeneous. This is true, with one exception—the multimeric complex. Take for example the giant molecule PDC. This is the pyruvate dehydrogenase complex, which catalyses three linked reactions, each on a separate enzyme protein. An apparently homogeneous preparation of this molecule may sprout 'impurities' when subjected to conditions that promote dissociation. These 'impurities', of course, are none other than the three constituent enzymes of the complex.

Putting these exceptions aside for the moment, let us assume that we have a homogeneous preparation of the macromolecule, and move on to studying its structure.

1.4.2 Molecular excluded volume

Macromolecules as studied by the biochemist are nearly always in solution. Because of their relatively enormous size they have a rather disrupting effect on the solvent molecules around them. This has to be taken into account when trying to make deductions about the size and shape of macromolecules from their behaviour in solution.

Consider what happens in aqueous solution when water molecules of molecular weight 18 come up against a small protein of molecular weight say 25 000. A molecule of this size will obviously disrupt the 3-D lattice structure of water (see Figure 6) over a large region, causing widespread interference with the free movement of water molecules, and it is this interference in solvent mobility that may be said to define a *non-ideal solution*. This is in contrast to an ideal solution where solvent movement is unimpeded. The precise extent of the macromolecular sphere of influence is what is known as the *molecular excluded volume*, MEV. This is the volume from which solvent molecules are *excluded*, and it is this, rather than molecular weight, that governs the experimental parameters we are interested in. Therefore, before these can be used to calculate molecular weight we have to fathom what factors other than actual molecular size are contributing to the MEV. Some of these are illustrated in Figure 10. One important factor is shape, and another is flexibility.* A long, thin, mobile molecule thrashing about in solution draws far more space under its influence than a compact, stationary one. This is why both shape and flexibility have to be considered when describing the behaviour of macromolecules in solution.

non-ideality

(a)

molecular excluded volume

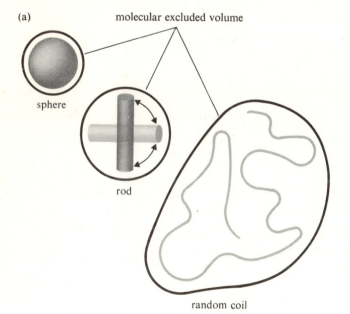

sphere

rod

random coil

(b)

molecular excluded volume

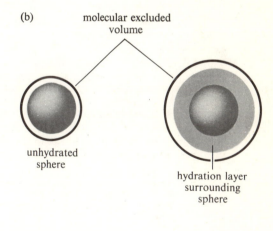

unhydrated sphere

hydration layer surrounding sphere

Figure 10 To show how (a) shape and flexibility, and (b) degree of hydration, affect the molecular excluded volume of a molecule. The masses of all five molecules shown are identical.

* A third important factor governing the sphere of influence is degree of hydration, i.e. the number of water molecules immobilized on the surface of the macromolecule (see Figure 10b). This tends to increase the core size of the molecule whose shape and flexibility we are concerned with, but for the sake of simplicity we have chosen throughout to neglect this aspect of the MEV. Unit 7 deals with hydration in more detail.

For our purposes there are three 'shapes' or conformations of molecules—spherical, rod-shaped and random coil. Most of the globular enzyme proteins are approximately spherical as the name suggests, while nucleic acids and many of the structural fibrous proteins and polysaccharides are rod-shaped. These are both native conformations, maintained by a high degree of secondary and tertiary structure, and their disruption of solvent mobility is usually assumed to be minimal, i.e. they can be taken to give ideal solutions. (Rod-shaped molecules, particularly long ones, sometimes strain this approximation to the limit, and special calculations are then needed.) The random coil which is produced on denaturation of either of the other two forms is rather different. It has lost all elements of secondary structure, has no unique 3-D shape and is a highly flexible molecule. Consequently it flays around in solution, disrupting the structure of the solute over a wide area, and it is this aspect that makes it the least ideal of the three 'shapes' in solution. In Figure 10 you can see what would happen to the MEV were a macromolecule of constant mass to change its shape from sphere to rod to random coil.

random coil

Perhaps molecular biology would be simpler if macromolecules did behave ideally all the time. Comparatively simple equations would then relate the measurable quantities (such as those in column 3 of Table 2) to the molecular weight function we are after. However, in trying to understand the reasons behind non-ideality it has been possible to turn this behaviour to advantage, using it to investigate shape as well as size from properties in solution.

1.4.3 How to deduce the shape of a macromolecule from its behaviour in solution

> **Study comment** In this Section we introduce three parameters commonly used to put the 'shape contribution' in numerical terms. Frictional coefficient is mentioned mainly because it is an important means of calculating D, the diffusion coefficient. Both this and intrinsic viscosity, $[\eta]$, are essential factors when it comes to calculating molecular weight from sedimentation velocity runs, and both will come up again in later Sections. The third parameter, radius of gyration (R_G), is included here because it is a particularly good way of visualizing the contribution of shape to MEV and may therefore illuminate the previous Section for you.

Radius of gyration In the previous Section on MEV we pointed out that the effective volume of a molecule is determined as much by its flexibility as by its actual molecular dimensions. The radius of gyration, R_G, is a good way of visualizing this. It is defined as the *effective radius* of a molecule, and intuitively would be expected to depend on flexibility. A rigid rod, for example, can hardly compete with a flexible coil of the same dimensions, which by thrashing around in solution can carve out a far greater effective radius for itself.

radius of gyration

> **ITQ 8** Consider a given protein molecule with a fixed molecular volume. Nothing is known about it shape, but in theory it may assume any one of three conformations, spherical, rod-shaped or random coil. Which of these three would you expect to give the smallest value for R_G?

Frictional coefficient A very useful way of discriminating between long, thin molecules and spherical ones is to measure their frictional coefficients. It is not hard to imagine that a particle being dragged through a solution under the influence of some external force, be it centrifugal or electrical, will experience opposition due to friction. Furthermore, this frictional resistance will be very dependent on shape—a compact spherical molecule will arouse less resistance than a thin, straggly rod for example. This frictional force between particle and surrounding water molecules is directly proportional to the frictional coefficient, f, and can be calculated very simply from the rate of diffusion, D. In D we come at last to something that can be measured experimentally. This is done by the (theoretically) simple means of finding the speed at which an initially sharp band of the macromolecule spreads into the surrounding medium. This operation is a very necessary one for proteins, where D, as we shall see later, is an essential piece of information for calculating molecular weight from sedimentation velocity runs.

frictional coefficient

diffusion coefficient

Intrinsic viscosity Finally we come to viscosity, which may be defined as the resistance of fluids to flow. The simplest viscometer is a long capillary tube along which the rate of flow can be timed. Everyone knows that water in such a tube would be expected to flow faster than say treacle, or even than a solution of sucrose. Again, intuitively, one would expect a strong sucrose solution to flow more slowly than a dilute one, i.e. viscosity is related to amount of solute. In fact, what counts is the now familiar molecular excluded volume, MEV. Molecules with a high MEV flow stickily, with a high viscosity, which is what you might have predicted from a brief look at Figure 10. What is done experimentally in the viscometer is to compare rates of capillary flow before and after addition of macromolecules to the solvent. Intrinsic viscosity, [η], which is a measure of the viscosity of the macromolecule itself, is then given by the relative flow rates of solvent and macromolecular solution.

intrinsic viscosity

This parameter, as we shall see later, is used to calculate molecular weight from sedimentation velocity data, but more interesting than this is the way it shows up *differences* in shape.

TABLE 3* To show how viscosity is influenced by shape rather than size

Macromolecule or particle	Shape	Molecular weight	Intrinsic viscosity/cm³ g⁻¹
ribonuclease	globular	13 700	3.4
bushy stunt virus	globular	10 700 000	3.4
myosin	rod-shaped	440 000	21.7
serum albumin (native)	globular	67 500	3.7
serum albumin (denatured)	random coil	67 500	52

* Modified from Table 7.1 of van Holde, K. E. (1971), *Physical Biochemistry*, Prentice Hall.

This is clearly shown in Table 3 where, as you will see, the different shapes adopted by native macromolecules or their multimeric complexes can readily be distinguished by their intrinsic viscosities. What is also clear from the Table is the overwhelming influence of shape rather than size. The rod-shaped myosin molecule, for example, has an intrinsic viscosity some six times greater than that of ribonuclease but this is obviously not a consequence of its greater size—the approximately spherical bushy stunt virus (a gigantic 10 millions) still has an intrinsic viscosity no bigger than that of ribonuclease.

Earlier in the Unit we mentioned the recent upsurge of interest in protein folding and unfolding, and it was with viscometry that some of the original work in this field was done. This is an example where the interest lies more in *change* of shape than in absolute dimensions.

ITQ 9 From what has been said about the MEV of spherical and random coil forms, what would you expect to happen to the viscosity of a globular protein on denaturation?

The effect of denaturation is shown quite convincingly in Figure 11, which shows how the protein ribonuclease is denatured by a rise in temperature.

ITQ 10 At what temperature would you say the protein in Figure 11 began to unfold from its native conformation?

Although we have talked about proteins rather than nucleic acids so far, it is perhaps with DNA that viscometry has done most for molecular biology. These long, thin molecules with their very high MEVs are particularly good subjects for the viscometer. The technique has not only been used to detect changes in shape (linear versus supercoiled, etc., see Figure 18, later) but, perhaps more important, it has also frequently been used as an empirical method of determining molecular weight, as described in the next Section.

NOW YOU COULD TRY SAQ 9.

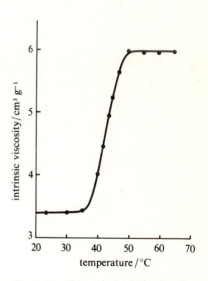

Figure 11 Change in intrinsic viscosity of ribonuclease during denaturation. The low value characteristic of a compact globular protein increases rapidly as heat treatment causes unfolding to the random coil form.

1.5 Popular empirical methods for measuring molecular weight

Nowadays, when it comes to measuring molecular weight in practice you will find that most biochemists turn if possible to an *empirical method*. This is one that requires calibration by standards of known molecular weight.

Calibration is needed because the method cannot stand on its own. The theory behind it is insufficiently understood to allow an unknown molecular weight to be calculated directly from experimental data on one compound. However, provided enough theory is known to predict conditions under which standard and unknown molecules behave in similar fashion, empirical methods are perfectly valid. One need only remember to keep within the rules, i.e. not to exceed the conditions for comparable behavior. This is more difficult than you might think, since such conditions are seldom very precisely understood.

It is only recently that sufficient standard molecules of known molecular weight have become freely available, so it is perhaps illogical to describe empirical methods before the 'absolute' methods which pioneered them. However, since our first method—viscometry—follows on from the previous Section, we shall ignore such illogicality and proceed.

In all three methods we are about to describe, you will find concern over one particular property of the molecules being examined. This concern is to ensure comparable behaviour of standards and unknown.

QUESTION From what we have already said about the behaviour of macro-molecules in solution, predict which molecular property this is.

ANSWER Shape. In what follows, note how each of the three methods tries to ensure uniformity of shape in both standards and unknown.

From an experimental point of view, most empirical methods are a great deal easier to handle than are absolute methods. However, there is no need to assimilate all the experimental details, only enough to understand what is going on and what precautions are being taken to ensure comparable behaviour of standards and unknown.

1.5.1 Viscometry and molecular weight

All that is measured here is intrinsic viscosity, as described in the previous Section. The unknown DNA, for example, is compared directly with standard DNAs whose molecular weights have been determined by light scattering (see Section 1.6). The shape difficulty is dealt with by simply assuming that all molecules have the same 'worm-like' coil configuration. The following empirical equation can then be applied,

$$[\eta] = aM^b \tag{1.4}$$

where $[\eta]$ is intrinsic viscosity, M the desired molecular weight, and a and b are constants. What the calibration runs with standards do is to provide values for a and b. Then M can be calculated directly from measurements of $[\eta]$.

1.5.2 Gel filtration

Gel filtration† has been described in an earlier Course[6]. The main point to recall is that large molecules pass rapidly through a gel of cross-linked poly-saccharide beads because they are excluded from all except the spaces outside the gel pores, whereas small molecules enter these pores, thus taking a more circuitous and longer route through the gel. What is measured is the *elution volume*, V_e. This is defined as the volume of fluid which needs to pass through a column of the gel before a sample applied to the top is washed completely through (see Figure 12a).

elution volume

ITQ 11 Would you expect V_e to be greater for (a) a small molecule or (b) a large molecule?

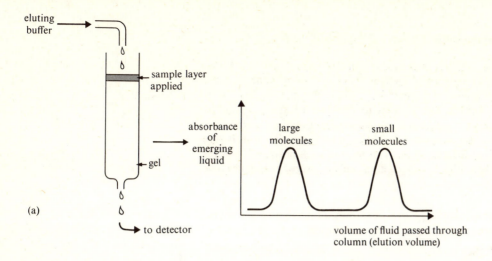

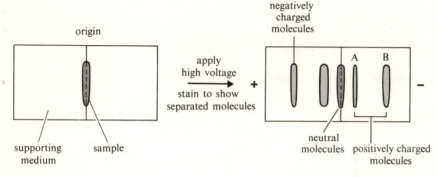

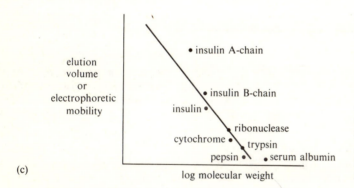

Figure 12 Gel filtration and electrophoresis. (a) Gel filtration. (b) Electrophoresis (a high voltage is applied across the supporting medium, and the separated molecules are then made visible by suitable staining; note that molecules A have a lower electrophoretic mobility than molecules B). (c) Relationship between molecular weight and parameter measured.

The main factor governing V_e is, as you might have guessed, molecular weight. Other factors like shape and degree of hydration can be discounted if we assume that both standards and unknown behave approximately as spheres* when passing through the gel. Some relationship such as equation 1.5 can then be used:

$$\log M \propto V_e \qquad (1.5)$$

One way of dealing with molecules that are obviously not spherical (or molecules that require dissociation before subunit size can be measured) is to convert everything—standards and unknown—to the same linear random coil shape. Running the column in a denaturing solvent such as **urea**** will do this, and the method is often used to find the size of individual polypeptide chains in a multimeric system such as haemoglobin.

denaturing solvents

* Strictly, non-hydrated spheres—see footnote to Section 1.4.2.

** Urea is a denaturing solvent, thought to act by disrupting H bonds in the macromolecule and its surrounding layer of water.

1.5.3 Polyacrylamide gel electrophoresis

Somewhat similar thinking lies behind another popular method, *polyacrylamide gel electrophoresis* in the presence of sodium dodecyl sulphate—the so-called 'SDS gels'. SDS is an anionic detergent* of formula $CH_3(CH_2)_{11}SO_3^-Na^+$. Its detergent action first reduces all proteins to the same linear shape, and its hydrophobic tail then binds to hydrophobic amino acid residues all along the length of the extended chain. A speculative diagram of this is shown in Figure 13. Since several hundred SDS molecules can bind to each chain, all proteins will end up, whatever their original composition, with an overwhelming negative charge.

SDS gels

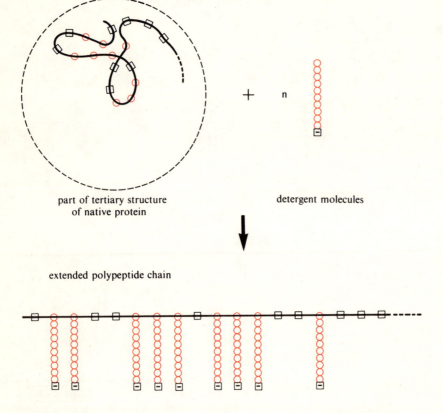

part of tertiary structure of native protein

detergent molecules

extended polypeptide chain

☐ denote polar residues

○ denote hydrophobic residues

Figure 13 Speculative diagram showing interaction between protein and sodium dodecyl sulphate (SDS) detergent.

Provided the SDS binding is not impaired at any point along the chain, the charge per unit mass should be constant. This is a very important point, because then the electrophoretic mobility (Figure 12c) will depend solely on molecular weight. A relationship similar to equation 1.5 can then be used, where V_e is replaced by *electrophoretic mobility*—the distance migrated by the protein relative to a marker dye.

electrophoretic mobility

$$\log M \propto \frac{1}{\text{electrophoretic mobility}} \qquad (1.6)$$

Not only proteins but also RNA molecules can be measured in this way.

1.6 Light scattering

We turn now to the absolute methods, which require no calibration and which have provided the 'standard molecules' on which the empirical methods depend.

Light scattering is an important technique that we shall mention only in passing. Perhaps its most significant application has been to provide a series of absolute molecular weights for the lower range of DNA molecules—those below a million —for use as standards. The technique depends upon the basic fact that visible

---

* A detergent, as will be described in Unit 3, is a molecule with polar residues in the head and non-polar (hydrophobic) residues in the tail.

light will be scattered when it passes through a solution of large particles—large in this context meaning particles that are still in true solution but whose dimensions are comparable to the wavelength of visible light. What is measured is the intensity of light scattered by the particles to various angles θ (theta) from the angle of incident light (see Figure 14). The scattered light is made to enter a photomultiplier. This is rotated to measure the intensity of the light scattered at a range of values for the angle θ.

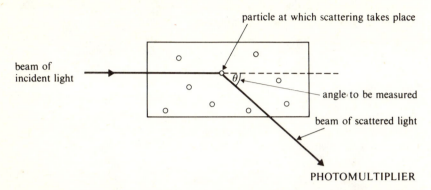

Figure 14   Light scattering.

The experimental data can also be manipulated to give another important characteristic of the macromolecule, its radius of gyration, R_G. This, as we said earlier, is a measure of the flexibility of the macromolecule in solution, and intuitively it is not surprising that this should have a strong influence on the intensity of light scattered to different angles.

1.7   The ultracentrifuge

At around £20 000 (in 1976) an ultracentrifuge is by no means cheap, but it is rare to find a research or analytical lab without access to one. The instrument can be put to various uses, as listed in Table 2 under sedimentation velocity and sedimentation equilibrium.

In *sedimentation velocity* the *rate* of sedimentation is measured; as you will see, this can be used to show up impurities in a preparation, to monitor dissociation of multimeric complexes or to characterize molecules and particles by their sedimentation coefficients. The method can be used for collecting samples, or for distinguishing between closely similar molecules if resolution is improved by replacing the buffer with a sucrose density gradient—a technique you will meet several times in this Course.

sedimentation velocity

The effect of shape on sedimentation rate is demonstrated by a digression into the tertiary structure of DNA. In some cases this shape contribution can be taken into account mathematically, so that molecular weight can be calculated. We have avoided details of the equations used, but you should be aware of which parameters can be measured during the velocity run and which, if any, have to be supplied from outside sources.

Sedimentation equilibrium is theoretically very different, and what is measured is the *position* of maximum concentration of the particles *at equilibrium*. For proteins this information can be used to calculate molecular weight, but for nucleic acids (whose long straggly molecules tend to break easily) such measurements are not feasible. What is measured instead is *buoyant density*, ρ_B (rho B) which requires quite a different system in which buffer is replaced by a caesium chloride (CsCl) gradient. Sedimentation in CsCl gradients is a remarkably sensitive technique, and the Section ends with an example of its powers of resolving closely similar molecules.

sedimentation equilibrium

1.7.1   Sedimentation velocity

Some people say that Svedberg is the father of molecular biology since it was he who first introduced the ultracentrifuge into biochemistry.* Whether or not this

---

* Svedberg and the ultracentrifuge are discussed in the radio programme *History of the macromolecule*.

is an exaggerated view you may be in a position to judge by the end of the Course, but certainly the scientific literature is full of references to physical constants which can only be obtained in the ultracentrifuge. This instrument is essentially a means of improving on gravity to produce centrifugal forces of up to $400\,000\,g$. Under these circumstances even one of the smaller protein molecules will be dragged, against the opposing forces of diffusion and friction, towards the bottom of the centrifuge tube or cell. A short time after the beginning of a run, therefore, there will be a *gradient of protein concentration* down the cell, as shown in Figure 15a. This shows the distribution of particles after t_1 seconds of centrifuging.

QUESTION From the information in Figure 15a, state which region of the centrifuge has the highest concentration of particles, that near the meniscus or that near the bottom.

ANSWER The region near the bottom of the cell is the position of maximum concentration, and particles can be seen piling up here. At the meniscus, concentration is at a minimum, tailing off to zero (i.e. pure solvent) as particles are removed from the region by sedimentation.

Somewhere in between these two extremes is the *boundary* between pure solvent and solution, and this is the point at which concentration is changing most rapidly. In Figure 15b, it can be seen as the steep part of the concentration versus distance plot. As sedimentation proceeds the boundary position moves slowly towards the bottom of the cell, and what is measured during a sedimentation velocity run is the *rate of change of boundary position*, i.e. dr/dt.

In many modern ultracentrifuges this rate can be measured directly, because the machines portray a concentration plot like that in Figure 15b. Boundary movement down the cell is followed by the progressive change in midpoint position r_1, r_2, etc. However, in most of the scientific literature cited here, boundary position has been followed indirectly using the classic Schlieren optics method. This measures the gradient of concentration as a *gradient in refractive index*[A]; a peak in the refractive index gradient corresponds to the position of maximum change in protein concentration, i.e. to the boundary position. Therefore in the Schlieren system what is followed is not the midpoint of a concentration plot, but the peak position in a plot of refractive index gradient. You can see this by comparing (b) and (c) in Figure 15. Because of diffusion the boundary becomes progressively broader and less sharp, and this is seen as a wider, shorter peak in the refractive index gradient. An example from the scientific literature is shown in Figure 16.

ITQ 12 Suppose after 80 minutes' centrifugation, an impurity in the preparation of the enzyme shown in Figure 16 becomes visible as a small peak, separating from the main enzyme peak on the meniscus side. What does this suggest about the relative molecular weights of enzyme and impurity? (Assume all molecules are approximately spherical.)

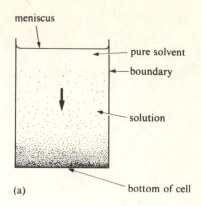

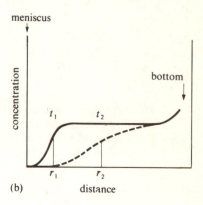

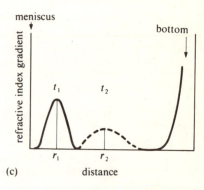

Figure 15 Movement of boundary position during a sedimentation velocity run. (a) Distribution of particles at a time t_1 seconds after centrifugation commenced. The arrow indicates sedimentation direction. (b) Direct method for following movement of boundary position—plot of concentration versus distance from centre of rotation. (c) Indirect (Schlieren) method for following movement of boundary position—plot of refractive index gradient versus distance from centre of rotation.

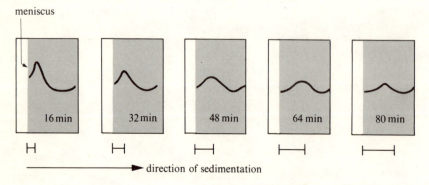

Figure 16 Sedimentation velocity run of purified methyl transferase, a key enzyme in the biosynthesis of adrenalin. The boundary, shown here as a peak in the Schlieren refractive index gradient, can be seen moving towards the bottom of the cell. Bars represent distance travelled by the boundary after different time intervals. (Data from Connett, R. J., and Kirshner, N. (1970), *J. Biol. Chem.* **245**, 331.)

Sedimentation coefficient (*s*) is expressed in Svedberg units (S), and is simply a measure of the rate of sedimentation per unit gravitational field. It is defined by:

$$s = \frac{\text{sedimentation rate}}{\text{gravitational field}} = \frac{dr/dt}{r\omega^2} \qquad (1.7)$$

sedimentation coefficient

where r is the distance from the centre of rotation, t the time and $r\omega^2$ is the angular velocity (measured in radians per second, rad s^{-1}) at which the rotor is turning. (The notation dr/dt is one way of expressing the rate of change of r with respect to t, i.e. the sedimentation rate.) Sedimentation coefficients are measured in units of $(\text{second})^{-1}$ and you will usually find them expressed as *Svedberg units*, S:

Svedberg unit

$$1\,S = 10^{-13}\,(\text{second})^{-1} \qquad (1.8)$$

The sedimentation coefficient may be used as it stands, without further elaboration. It is simply one way of characterizing a given particle or macromolecule, and for this there is no need to worry about the relative contributions of size and shape to the final S value. Remember the 50S and 30S subunits of the ribosome in Figure 4. Here is an example where nothing need be known about structure or function, but the particle is labelled and identified for future reference by its S value. The Case Study on pyruvate dehydrogenase (see Section 1.8) has some more examples of this.

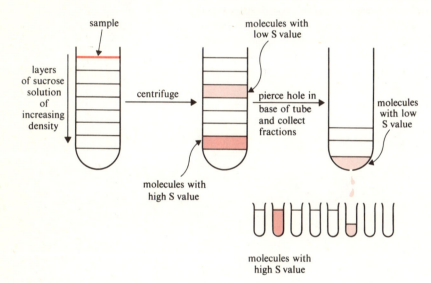

Figure 17 Sucrose density gradient centrifugation.

Very often it is desirable to be able to collect and analyse the separated macromolecules. This would be difficult in the set-up we have just described because the particles are distributed in a concentration gradient throughout the cell. For preparative work the particles can be made to sediment in a band, by centrifuging not in dilute aqueous buffer but in a *sucrose density gradient*. Here the sample does not start off as a homogeneous solution filling the cell in which sedimentation takes place, but is layered as a 'zone' on top of a sucrose density gradient already in the cell. This is shown in Figure 17, where you can see that the gradient consists of a series of sucrose solutions of increasing concentration, with the densest (most concentrated) one at the bottom. The particle, being denser than even the bottom sucrose layer, will tend to sediment slowly down the cell, and it is this process that is greatly accelerated by centrifugation. As before, the rate of sedimentation depends on molecular size and shape, but in the gradient these differences are exaggerated so that even closely similar molecules will separate. Furthermore, the particles move as a band which does not rapidly disperse when the rotor stops, but can be collected for further analysis, as shown on Figure 17.

sucrose density gradient centrifugation

In the great majority of published papers concerning the role of RNA in protein synthesis, you will find mention of this technique—sedimentation velocity in sucrose density gradients. Its importance in this field cannot be overstated. The cell has a large variety of RNA molecules, and many whose function is still obscure have been well characterized by their S values. You will hear more of this in Unit 11.

So far we have described how sedimentation velocity runs, whether with or without sucrose density gradients, can be used to determine S values. However, far more information can be obtained if we consider in greater detail what it is that makes a molecule sediment at a particular speed. It is obviously something to do with mass, i.e. molecular weight, but the two are not directly proportional. This is because molecules of identical mass may not have the same shape and, as we said in Section 1.4.2, this means they will have different MEVs and therefore behave differently in solution. As far as sedimentation goes, the shape contribution is very dramatically seen in the behaviour of three DNA molecules—all of which have the same nucleotide sequence but different shapes. Because this relationship between primary and tertiary structure in nucleic acids is somewhat contrary to current thinking on proteins, we shall digress at this point into a discussion of the tertiary structure of DNA. (This discussion also provides an opportunity to point out that the ultracentrifuge has a role in solving problems in biology, and is not just another biophysical technique.)

1.7.2 Tertiary structure of DNA elucidated by sedimentation velocity

We know far less about the tertiary structure of DNA than we do about that of proteins or even tRNA, but three distinct forms of DNA can be recognized. These are all variations on the theme of the double helix, and in theory may be derived from one another as follows. The open-cyclic form (Figure 18) may be visualized as the linear form covalently joined at the ends to form an open circle. To make the superhelical form, imagine that one chain of the double helix is cut, unwound a few turns and quickly resealed. This causes the whole molecule to buckle into 'supercoils' in an effort to rewind and revert to the original structure. The three forms as they occur in a cell extract may be distinguished—as you might have guessed—by their behaviour in the ultracentrifuge. The linear molecule is the most flexible of the three, while the supercoiled helix is the most compact and dense.

> **ITQ 13** From your knowledge of the effect of shape and flexibility on the MEV (molecular excluded volume) of macromolecules in solution, how would you arrange solutions of the linear, open-cyclic and superhelical forms of DNA in order of decreasing sedimentation rate?

In this way three different forms of the same DNA molecule can be distinguished by their behaviour in the ultracentrifuge—and it is this kind of experiment that has been behind much of our recent understanding of DNA tertiary structure. Unfortunately, as yet there is little to relate structure to biological role but it is not hard to imagine that some forms may be more accessible to outside agents, and therefore possibly less inert, than other forms. For example, tightly coiled superhelical DNA may be difficult to transcribe into RNA. Most of the present work relates to the more manageable DNA of viruses, mitochondria, etc.; the tertiary structure of the enormous stretches of chromosomal DNA found in higher organisms is much less well understood.

NOW YOU COULD TRY SAQ 10.

We return now from nucleic acids to the theory behind sedimentation velocity runs. If we can put some exact figures on the shape contribution, sedimentation rate can be used to measure molecular weight. As we have just said, long thin molecules will sediment more slowly than spherical ones, even where the masses are identical. In order to eliminate this sort of anomaly, molecular weight calculations therefore include the term D, the diffusion coefficient. As we saw in Section 1.4.3, D puts a figure on the shape factor. Once D is known, sedimentation rates can be used to calculate molecular weight. This is a procedure which has been carried out many times for proteins, but nucleic acids present more of a problem since D is not easily measured for such long thin molecules. However, for nucleic acids the intrinsic viscosity, $[\eta]$, will do instead.

We have now shown how sedimentation rates can be used in three ways: to distinguish between molecules by giving them characteristic S values, to differentiate between molecules of identical size but different shape, and finally (given extra information in the form of $[\eta]$, D or standards of known molecular weight) to calculate molecular weight. All these measurements are made from sedimentation velocity runs. Sophistication in the form of sucrose density gradients for extra resolution may or may not be present, but in all cases the system

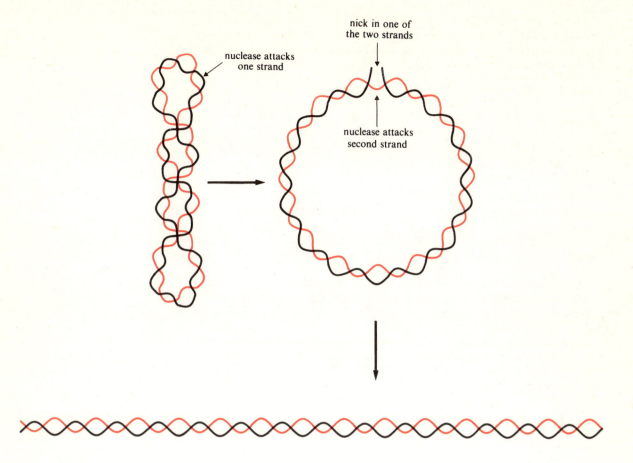

is far from being at equilibrium. Sedimentation proceeds rapidly in the face of diffusion and friction, and the main experimental task is to measure s, the sedimentation rate.

Figure 18 Tertiary structure of DNA.

1.7.3 Sedimentation equilibrium

With sedimentation equilibrium the situation is quite different. The ultracentrifuge is allowed to run much more slowly, and eventually an equilibrium is reached, in which the position of each particle remains almost constant. At this point the centrifugal force on each particle is exactly balanced by the opposing forces of friction and diffusion. The particles will then be found distributed in an *equilibrium gradient of concentration*, with the highest concentration at the bottom of the cell. A great advantage of using sedimentation equilibrium to determine molecular weight is that all the necessary information can be found directly from the ultracentrifuge run, by analysing this concentration gradient. There is no need to rely on outside information such as D or $[\eta]$, or to compensate for any differences in shape, since at equilibrium these will cancel out. This is the advantage of centrifuging to the point at which upward and downward forces on the particle are equal.

Sedimentation equilibrium in this form (i.e. without a density gradient) is a highly accurate method for determining the molecular weight of proteins, and has been of major importance in providing absolute values for calibrating other methods. It is of little use for nucleic acids, since these tend to break on handling and give erroneously low molecular weight values. However, provided the original nucleic acid is fairly homogeneous along its length (i.e. no clustering of adenine and thymine base pairs $(A+T)$ or of guanine and cytosine base pairs $(G+C)$ at either end) then the nucleic acid fragments will retain the buoyant density ρ_B of their parent. This parameter, ρ_B, varies widely with nucleic acids from different sources and is a very useful diagnostic feature. (Globular proteins, on the other hand, tend to have rather similar densities and cannot readily be distinguished by density measurements.) Buoyant density can be determined in a sedimentation equilibrium run by using a density gradient similar to that described for preparative sedimentation velocity runs. But here the gradient is produced not by sucrose solutions, which are always less dense than the

buoyant density

macromolecule, but by caesium chloride solutions. These provide a range of solutions whose density encompasses that of the macromolecule. Therefore at equilibrium the macromolecule should be concentrated in a band, at a position where its own buoyant density will equal the density of the CsCl at that point.

The resolving power of this technique is remarkable. Buoyant density in nucleic acid chemistry has been extensively used to differentiate between molecules of identical composition but different shape, to pick out similar DNAs from different sources, and finally, most remarkable of all, to distinguish between DNA molecules identical in both sequence and shape but with different isotopic content, e.g. those where one or both strands of the double helix contain ^{15}N in place of the normal ^{14}N. This last feat refers to the famous Meselson and Stahl experiment to demonstrate semi-conservative replication in DNA. This was described in some detail in an earlier Course[7], and we shall now give you some of the experimental evidence.

Meselson and Stahl experiment

At every cell division, hereditary material in the form of DNA is divided in such a way that each daughter cell receives an exact copy of the original. This is the process of *replication*, which in theory may take one of at least two forms, as shown in Figure 19. In conservative replication one daughter receives both strands of the original while the other has both the newly synthesized strands, while in semi-conservative replication each daughter receives one original and one newly synthesized strand.

DNA replication

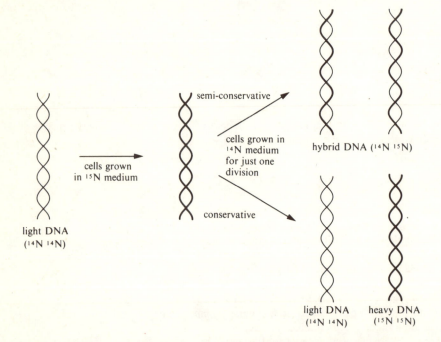

Figure 19 Alternative mechanisms for replication of DNA.

Before we could begin to understand just how DNA acts as a template for such exact replication, it was essential to be able to distinguish between these two alternatives.

Cells with double-stranded 'heavy' DNA (containing ^{15}N in place of the normal ^{14}N) were prepared by growing *E. coli* bacteria on a medium in which all the available nitrogen was in the form of the heavy isotope, ^{15}N. We shall describe the heavy double-stranded DNA isolated from these cells as $^{15}N\,^{15}N$, indicating that both strands carry heavy isotope. Such DNA could be distinguished from normal 'light' DNA ($^{14}N\,^{14}N$)—isolated from cells on normal growth medium —only by using the most sensitive of ultracentrifugation techniques, sedimentation equilibrium in a caesium chloride density gradient. You can see the results in Figure 20b, which shows the density gradient pattern that results when these heavy cells are returned to normal ^{14}N growth medium just long enough for one cell division to take place.

At this stage you are expected to do no more than appreciate the role of the ultracentrifuge in separating such remarkably similar macromolecules.

For the contribution of this famous experiment to molecular biology, turn to SAQs 11 and 12.

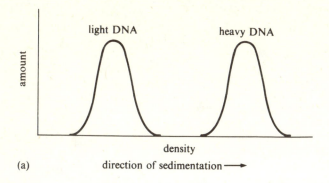

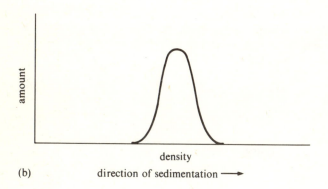

Figure 20 Use of the ultracentrifuge to determine the mechanism of DNA replication. (a) Calibration run in caesium chloride density gradient using pure heavy and pure light DNA. (b) Sedimentation of DNA from cells transferred from heavy to light growth medium for one cycle of cell division.

1.8 Case Study on pyruvate dehydrogenase complex (PDC)

Although our survey of techniques is not yet finished we shall break off here and attempt to justify the energy you have spent so far, by means of a Case Study. This concerns the structure of PDC—a giant multimeric complex (molecular weight 5 million) catalysing three consecutive enzyme reactions. How would you set about tackling such a problem? The research work described here relied heavily on the ultracentrifuge. However, it is very seldom that a single approach can provide all the answers to a biological problem, and a second quite different technique—electron microscopy—was also used. We shall describe this work too, even though the electron microscope is not formally introduced until Section 1.9.

The present Section begins by setting the picture with a description of the overall enzyme reaction and the steps catalysed by individual components. It then explains the necessary preliminary work, in which special chromatographic techniques were devised to separate each enzyme out from the complex, and special enzyme assay conditions were devised to substitute for reactants normally provided by other members of the complex. Sedimentation velocity data and electron microscopy then show how the complex may be reconstituted.

1.8.1 Measuring the enzymic activity of PDC and its component parts

Pyruvate dehydrogenase is a multimeric complex which consists of three enzymes —a decarboxylase (Ea), a transacetylase (Eb) and a dehydrogenase (Ec). Each catalyses one step in a series of reactions which result in the oxidative decarboxylation of pyruvic acid. In this way the enzyme is responsible for channelling pyruvic acid from the end of glycolysis† into the citric acid cycle.† Its central role in metabolism therefore hardly needs emphasizing, but it is also worth studying as a model for multimeric complexes in general. During evolution, each enzyme must have 'taken the trouble' to evolve a specific binding site not only for its own substrate but also for its adjacent enzyme, Ea, Eb, or Ec. In this way the active sites of sequentially-acting enzymes are brought close enough for substrate to be passed from one to the other without diffusing off into free solution. This is thought to make a multimeric complex more efficient than the free-floating enzymes of, say, the glycolytic pathway.

PDC structure

The overall reaction catalysed by PDC is:

$$CH_3-\underset{\underset{O}{\|}}{C}-COO^- + CoA \xrightarrow[\underset{NAD \quad NADH_2}{}]{lipoic\ acid,\ FAD,\ TPP} CH_3-\underset{\underset{O}{\|}}{C}-S-CoA + CO_2 \qquad (1.9)$$

pyruvate   acetyl-CoA

Lipoic acid, TPP and FAD do not appear in the overall scheme but are required as cofactors or coenzymes.† Their precise role can be seen from the breakdown into steps 1–5 (Figure 21). In step 1, Ea decarboxylates pyruvate, releasing CO_2 and leaving the group that is shown in the margin on the right; this group is bound to Ea as a covalent ES intermediate (see Unit 5). In step 2 this group is transferred to the lipoic acid moiety that is covalently attached to Eb, the trans-acetylase. Also bound (by non-covalent forces) to the active site of Eb is the second substrate of the overall reaction, coenzyme A (CoA).† In step 3 this receives the acetyl group from the lipoic acid and is released into the medium as acetyl-CoA, the major product. However, the lipoic acid is still in a reduced form and the two remaining steps are concerned with its re-oxidation. In step 4 it is oxidized at the expense of an FAD molecule bound to the active site of Ec, the dehydrogenase. In the final step this $FADH_2$ moiety is re-oxidized by reaction with NAD, releasing $NADH_2$ into the medium.

$$\boxed{pyruvate} + \text{Ea-TPP-H} \quad\rightleftharpoons\quad \boxed{CO_2} + \text{Ea-TPP-CHOH-CH}_3$$
(step 1)

$$\text{Ea-TPP-CHOH-CH}_3 + \text{lipoyl-Eb} \rightleftharpoons \text{Ea} + \text{acetyl-lipoyl-Eb} \quad \text{(step 2)}$$

$$\text{acetyl-lipoyl-Eb} + \text{CoA} \quad\rightleftharpoons\quad \text{acetyl-CoA} + \text{dihydro-lipoyl-Eb}$$
(step 3)

$$\text{dihydro-lipoyl-Eb} + \text{Ec-FAD} \quad\rightleftharpoons\quad \text{lipoyl-Eb} + \text{Ec-FADH}_2$$
(step 4)

$$\text{Ec-FADH}_2 + \boxed{NAD} \quad\rightleftharpoons\quad \text{Ec-FAD} + \boxed{NADH_2} \quad \text{(step 5)}$$

Figure 21 Reactions catalysed by pyruvate dehydrogenase complex. Red shading indicates reactants that appear in the overall reaction. The other reactants are passed rapidly from one active site to the other, and are not released into the medium. Formulae of coenzymes NAD, FAD and TPP can be found in the *Source Book*.

ITQ 14 From the information in equation 1.9, suggest a means of measuring the activity of the intact complex.

In practice you will find that many enzyme assays are built around the very convenient change in absorbance† at 340 nm, which is the absorption maximum† for $NADH_2$. If we use pyruvate as the substrate and incubate it with NAD plus *E. coli* extract, then any rise in light absorption at 340 nm should mean $NADH_2$ is being produced by oxidation of pyruvate, i.e. that the PDC complex in the extract is working as a whole. Using this assay, the enzyme activity was pursued through several purification steps and the homogeneity of the end-product was checked by a sedimentation velocity run. This revealed a single peak of 57S. Such a high S value already aroused suspicions that it was a multi-component system, and these were confirmed on finding that if this preparation were chromatographed at high pH, the 57S component split into two fractions, a small colourless fraction of 9S (see Fig. 22) and a large yellow component of 30S. (Note the use of S values to characterize a component: even if nothing else is known about it the protein may be recognized each time from this reproducible physical measurement. Note also that the S values are not additive, i.e. 9 plus 30 does not give 57!). The actual identity of these two fractions, the colourless and the yellow, could be deduced from other information such as the following:

1 The colourless 9S component contained no lipoic acid or FAD, and catalysed the following reaction:

$$\text{pyruvate} + Fe(CN)_6^{3-} \longrightarrow \text{acetate} + CO_2 + Fe(CN)_6^{4-} \quad (1.10)$$

This is essentially the same as steps (1+2) in that pyruvate is decarboxylated, but differs in detail because Eb has been replaced by ferricyanide. (The reaction was followed by monitoring the rate of ferricyanide reduction.)

2 The yellow 30S component contained all the lipoic acid and FAD of the original PDC complex.

assay of intact PDC

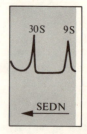

Figure 22 Sedimentation velocity run of fractions obtained by chromatography of PDC at high pH. The run was carried out in a buffer which prevented re-association. The apparently homogeneous enzyme split into fraction Ea (9S) and a partially dissociated complex of Eb+Ec (30S).

assay of Ea

40

ITQ 15   What does this tell you about the probable locations of subunits Ea, Eb and Ec within the two fractions?

The answer to ITQ 15 was confirmed by further experiments in which the yellow component was chromatographed in the presence of 4 M urea. Two smaller fractions resulted, with the catalytic activities to be expected of Eb and Ec respectively. (Another modified enzyme assay was devised like the ferricyanide reaction (equation 1.10) for Ea, to substitute for residues normally provided by the other enzymes of the complex.)

dissociation of Eb–Ec complex

You may be puzzled by the use of urea as a denaturant. This is thought to act by disrupting H bonds in the macromolecule and surrounding water. With care, in the case of PDC as described here, one can arrange for it to disrupt only the most exposed bonds (i.e. those maintaining quaternary structure), leaving secondary and tertiary structure intact. This dissociates the multimeric complex without denaturing its component parts. (High pH was used for the same purpose, in chromatography of the original 57S complex, as just described.)

ITQ 16   Suggest what precautions could be taken to guard against denaturation of components Ea, Eb and Ec during dissociation of the complex by urea or high pH.

1.8.2   Reconstitution of PDC from isolated preparations of its component parts

We have now established that PDC contains three types of enzyme each catalysing an individual step in the overall sequence. Looking at the reaction sequence in Figure 21, what considerations would you expect to govern the parcelling together of Ea, Eb and Ec? Eb obviously has a central role, in that the lipoic acid covalently bound to its active site has to interact both with Ea (step 2) and then with the FAD in the active site of Ec (step 4). There is evidence, which we shall now describe, to suggest that Eb is the most highly specialized of the three, in that it binds specifically to both Ea and Ec. Ea and Ec, on the other hand, have no affinity for each other. How do we know all this? Again the answer lies in the ultracentrifuge (see Figure 23). The three individual components, purified by chromatography in alkali or in urea, were mixed together in the same proportions by weight in which they occurred in the original complex. The sedimentation coefficients of these mixtures are shown in Table 4. What we have not shown

specific binding sites

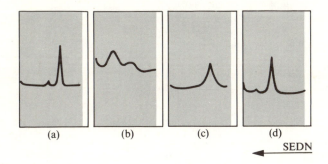

(a)   (b)   (c)   (d)

← SEDN

Figure 23   Sedimentation velocity runs to illustrate reconstitution of PDC after dissociation into enzymes Ea, Eb and Ec. (a) Fully reconstituted complex appears as a single peak. (b) Lack of interaction between Ea and Ec, which sediment as two separate peaks. (c), (d) Interaction between Ea and Eb (in c) or Eb and Ec (in d). When mixed in equimolar proportions, each pair sediments as a single peak.

TABLE 4   Reconstitution of PDC in the ultracentrifuge

	Enzyme	Number of peaks in sedimentation velocity run	Sedimentation coefficient
1	Ea	1	9S
2	Eb	1	24S
3	Ec	1	6S
4	Ea+Eb+Ec	1	57S
5	original PDC	1	57S
6	Eb+Ea	1	50S
7	Eb+Ec	1	30S
8	Ea+Ec	2	9S+6S

is the accompanying enzyme activities, and these may be summarized as follows:

1   Each individual component retained its activity when separated out from the complex.

2   When individual components were recombined to give 'reconstituted PDC', they again collaborated to catalyse the NAD-linked oxidation of pyruvate—just as in the original complex.

3   Partially reconstituted PDC, such as Ea+Eb, or Eb+Ec, showed the expected pairs of enzyme activities.

These two lines of evidence, sedimentation velocity and enzyme assays, together confirm that Eb can combine with both Ea and Ec, and that Ea and Ec do not interact except through their common attachment to Eb.

NOW YOU COULD TRY SAQ 13 TO CHECK THAT YOU HAVE UNDERSTOOD THE CONTRIBUTION OF SEDIMENTATION RUNS TO THIS CONCLUSION.

At this stage we should mention that the picture is even more complex than it may have appeared so far. Each of the three enzyme components occurs more than once in the PDC molecule, and furthermore each is itself multimeric. For example, mild dissociation of PDC produces not one but eight molecules of Eb, each of which is enzymically active. Under more stringent conditions each of these enzyme molecules dissociates further into three inactive subunits. A subunit, remember, is a single polypeptide chain and represents the point of maximum dissociation.

TABLE 5   Tentative grouping of Ea, Eb and Ec subunits in PDC

| | | | Subunit grouping | |
Enzyme	Molecular weight of subunit	Total number of subunits in one molecule of PDC	Number of subunits per active enzyme group	Number of active enzyme groups per PDC molecule
Ea	90 000	24	2	12
Eb	36 000	24	3	8
Ec	56 000	24 (12)*	2	12 (6)*

* The figures in parentheses for Ec take account of recent data, see Figure 24g.

Table 5 summarizes the subunit grouping for all three types of enzyme. As you can see, a single PDC molecule is now thought to contain 24 subunits of each of the three enzymes. Because these are clustered in groups rather than combined in the ratio 1:1:1, the quaternary structure of PDC is still written as $Ea_{12} Eb_{24} Ec_6$. The data in Table 5 have been very controversial and exceedingly difficult to establish* (some textbooks still carry outdated figures). A major problem has been persuading each active enzyme group to dissociate fully, so that the number and size of its component subunits could be determined.

ITQ 17   Suggest (a) conditions suitable for completely dissociating the Eb enzyme, and (b) a method of determining the molecular weight of the Eb subunits produced by this dissociation.

1.8.3   The final picture of the complex

We shall now focus on the complex more closely, in an attempt to see how all these active enzyme groups are arranged for maximum efficiency of the complex as a whole. The main criterion is to allow Eb easy access to the active sites of both Ea and Ec. At this stage it is appropriate to bring in the electron microscope, even though a detailed discussion of the technique is deferred till Section 1.9. The individual components and the various permutations of partially reconstituted PDC (e.g. Ea plus Eb) when viewed under the electron microscope give the results shown in Figure 24. As you can see, interpretation is by no means

* Compare Figure 24f and g.

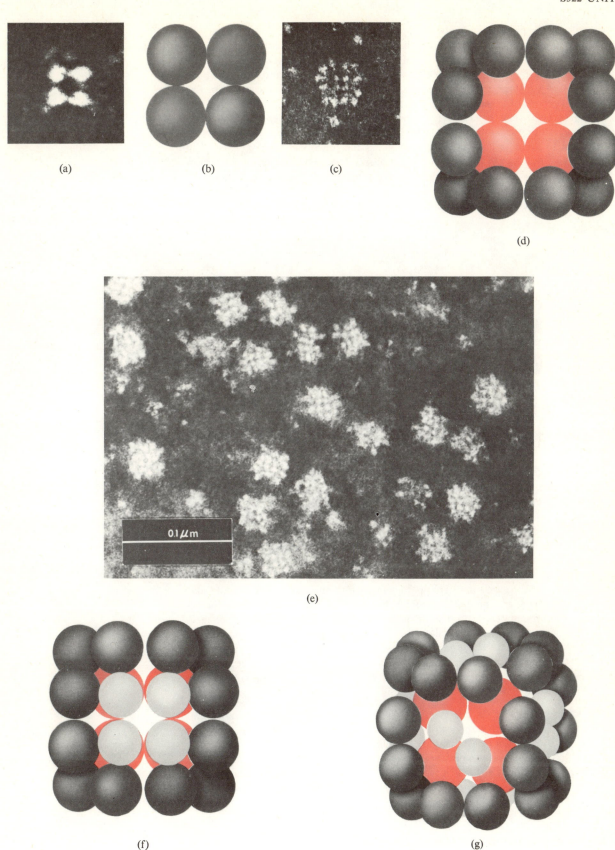

Figure 24 Electron micrographs and interpretative models of native and partially dissociated PDC. (a) Electron micrograph of the Eb core ($\times 870\,000$ magnification). The eight Eb enzymes are thought to be arranged at the eight vertices of a cube; note that only four enzymes are visible in this end-on view. (b) Interpretative model of (a) (subunits of Eb are not shown). (c) Electron micrograph of Ea–Eb complex; same view as (a), at only $\times 200\,000$ magnification. The Eb core can be seen surrounded by 12 of the 24 Ea enzymes. (d) Interpretative model of (c) (black spheres are Ea, red spheres are Eb; subunits are not shown). (e) Electron micrograph of native PDC, showing complete complex from different angles ($\times 300\,000$). (f) Interpretative model of (e) showing how Ec (small white spheres) would fit onto Eb core (just visible as red spheres) were it to occupy all the available space. (g) Recent interpretative model of (e), incorporating data which suggest that (i) not all the available space can be simultaneously occupied, and (ii) there are only 12, not 24, Ec subunits, and these are arranged in groups of 2, not groups of 4 (data from L. J. Reed *et al.* (1975), *Proc. Nat. Acad. Sci.* **72**, 3068).

43

easy. An inspired guess can be fortified by the information in Table 5, which tells just how many black, white or red balls there are to fit in. (Remember that each ball contains several subunits, but this subdivision is not shown except in the case of Ec.) Of the various interpretative models of the electron micrographs we have selected just three, to emphasize how the whole complex is built around Eb. The eight groups of Eb subunits are thought to be arranged at the eight vertices of a cube. The twelve pairs of Ea subunits then fit around this, leaving spaces into which fit the six groups of Ec subunits. This is a structural model, but what we want to know is how far it accommodates the functional requirements of an enzyme that has to proceed through reactions 1–5, and whether all the appropriate active sites are in the right place for substrate to pass directly from one to the other. Figure 25 summarizes all the available data in terms of a 'functional complex' that appears to fit both structural and enzymic requirements.

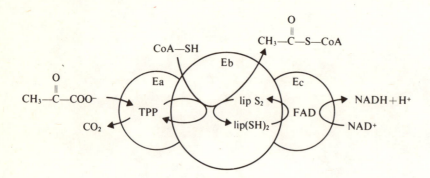

Figure 25 Diagram of the 'functional complex' of PDC, showing juxta-positioning of active sites of Ea, Eb and Ec.

1.9 Microscopy

In this final Section we complete our survey of techniques by describing the kind of information that comes from looking 'directly' at macromolecules. This is quite a different approach from all the indirect methods we have described so far. X-ray diffraction, which operates right down to 0.1 nm resolution, is mentioned only in passing as it comes up again in Units 2 and 3, and most of the examples here are from electron or light microscopy. The electron microscope can resolve objects 1 nm apart and is quite capable of revealing the dimensions of even a fairly small protein. (This would be about 2 nm across, i.e. 1/500 the width of a mitochondrion.) The instrument can give information on the molecular weight, shape and subunit structure of proteins. You have already had one example of this in PDC, a protein with globular subunits, and here we add collagen, a structural protein composed of fibrous subunits. The electron microscope has also made a great impact on our understanding of the structure and function of DNA, but it does have one area of uncertainty. One has to assume that the dried-out samples prepared for microscopy have not lost their native hydrated structure to any great extent.

Some nucleic acids are large enough to be seen in the light microscope, and either type of microscopy can be combined with autoradiography to provide a valuable means of measuring the molecular weight of very large DNA molecules. At the same time it can give a picture of biology in action at the molecular level —as is shown by the classic autoradiograph of *E. coli* DNA replication. The Unit ends by emphasizing just how hazy is the borderline between macromolecule and cell organelle.

1.9.1 Protein structure and the electron microscope

We begin with some examples of the use of the electron microscope (EM) in protein studies. Determining molecular weight is, in theory, very simple. All that is needed is to count the number of protein particles in a given field, and to measure the concentration of protein in the solution from which the slide was prepared. But perhaps the electron microscope has really come into its own in deciphering subunit structure in the larger proteins. Take a fibrous protein such as collagen (Figure 26). The basic units are tropocollagen molecules which pack together into fibrils. These can just be made out in the EM as striations in the collagen fibres and sheets. (The fibre arrangement of collagen is 'designed' for

collagen

(a)

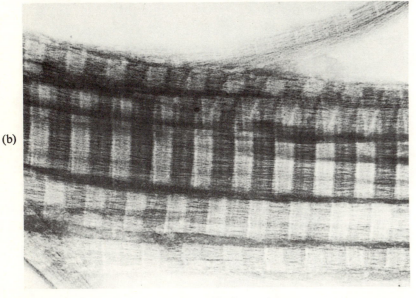

(b)

Figure 26 Structure of collagen. (a) Electron micrograph at ×30 000 magnification. (b) Electron micrograph at ×300 000 magnification, showing individual fibrils. (c) Diagram from X-ray diffraction data showing part of the triple-helix structure of individual molecules.

(c)

stretching, as in tendon, whereas the sheets are 'designed' for protection, as in cornea.) The tropocollagen building block is already a large molecule as compared with a simple globular enzyme like ribonuclease (molecular weights 300 000 and 14 000 respectively) and is itself composed of a well-defined triple helix. All structures below the fibril level, including individual tropocollagen molecules, are unfortunately too small to be resolved by the EM, and this is the point at which X-ray diffraction takes over. The different hierarchial levels of collagen structure are shown in Figure 26; (a) and (b) are within the scope of microscopy, while (c) comes from X-ray diffraction data.

Electron microscopy has also been used to study subunit structure when the building blocks are globular rather than fibrous proteins, e.g. the multi-enzyme complex, pyruvate dehydrogenase, just described.

1.9.2 DNA and autoradiography

The molecular weight of large DNA molecules is difficult to measure by the indirect methods we described earlier, being around 1–100 million for viral DNA, for example (this makes it some 10 000 times the size of an average protein like ribonuclease). What is known with some accuracy from X-ray work is the molecular width and the mass per unit length. The width of double stranded DNA is around 2 nm, bringing it just within the limits of resolution of the EM. It should, therefore, be possible to measure out the entire length of a DNA molecule from an electron micrograph and hence calculate molecular weight from our second piece of data, the mass per unit length. The molecule becomes easier to trace if its narrow width is amplified some 500-fold by *autoradiography*. Radioactively labelled DNA is placed in contact with a photographic plate, which after several weeks' exposure shows a continuous line of exposed silver grains along the length of the molecules. One difficulty of the technique is to persuade the molecule to lie flat and untangled during fixation, but Figure 27 shows one classic example where this has been achieved. A bacterial chromosome of molecular weight 3 200 million is shown during replication. This autoradiograph not only shows one of the largest molecules whose molecular weight is known with any accuracy, but also demonstrates one of the basic processes of biology, namely DNA replication, at the molecular level.

NOW YOU COULD TRY SAQ 14.

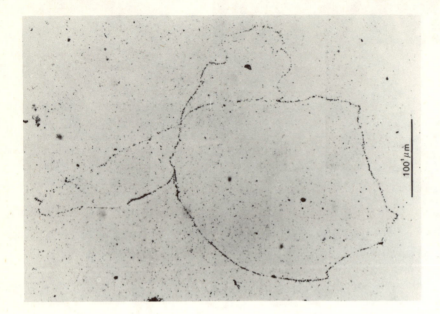

100 μm

Figure 27 Autoradiograph of *E. coli* chromosome during replication. This shows a single DNA molecule of molecular weight 3×10^9. The double stranded region which has already been replicated is on the right-hand side.

1.9.3 Organelles

In dealing with collagen fibrils and bacterial chromosomes we have crossed the borderline between macromolecule and cell organelle. Where exactly does this borderline fall? A simple virus such as tobacco mosaic virus, which is composed of 2 100 protein subunits and one RNA molecule, has an overall weight of 40×10^6. Would you classify this as a macromolecule of quaternary structure, $protein_{2\,100} RNA_1$, or as an organelle? What would you say about the pyruvate dehydrogenase complex of quaternary structure $Ea_{12} Eb_{24} Ec_6$ and molecular weight 5×10^6?

There is no cut-and-dried answer here, although viruses are usually thought of as supramolecular structures in the same category as ribosomes and mitochondria, while PDC is a rather large molecule. The important point to remember is that all levels of organization—organelles, cells, tissues and whole organisms —are made up of macromolecules. In the final analysis, the whole of biology with its interactions between cells, organisms, and even species, can be seen as an extension of interactions at the molecular level. This is why we have considered it important to spend some time discussing the structure and properties of macromolecules.

References to other Open University Courses

1 S2–1, Unit 1, Section 1.5

2 S2–1, Unit 1, Section 1.4.4

3 S2–1, Unit 1, Section 1.4–1.7

4 S2–1, Unit 1, Section 1.6

5 S2–1, Unit 6, Sections 6.3, 6.4

6 S2–1, Unit 1, Section 1.2.1

7 S100, Unit 17, Section 17.2.2

8 S2–1, Unit 2, Section 2.2

Recommended reading

LEHNINGER, A. L. (1975), *Biochemistry*, 2nd edn, Worth Publishers Inc.
A very good if expensive book. Chapters 2 and 6 are relevant to this Unit.

MAHLER, H. R. and CORDES, E. H. (1971), *Biological Chemistry*, 2nd edn, Harper and Row.
Chapters 3 and 5 cover Part II of the Unit very well.

STACY, G. W. (1975), *Organic Chemistry: A Background for the Life Sciences*, Harper and Row.
Contains the chemistry essential for this Course.

VAN HOLDE, K. E. (1971), *Physical Biochemistry*, Prentice-Hall Inc. (Foundations of Modern Biochemistry Series.)
More advanced treatment of the theory behind the techniques described in Part II.

WATSON, J. D. (1975), *Molecular Biology of the Gene*, 3rd edn, W. A. Benjamin Inc.
An excellent book. Chapter 4 is particularly relevant to this Unit.

WOLD, F. (1971), *Macromolecules: Structure and Function*, Prentice-Hall Inc. (Foundations of Modern Biochemistry Series).

YUDKIN, M. and OFFORD, R. (1973), *Comprehensible Biochemistry*, Longmans.
Another excellent book. Chapters 3 and 4 are particularly relevant to this Unit.

Self-assessment questions

SAQ 1 (*Objective 1*)   Write down the formulae of the disaccharide or dipeptide formed when the following pairs of building blocks combine by condensation:

(a) β-*N*-acetylglucosamine+β-*N*-acetylglucosamine

(b) glycine+alanine

(c) glutamic acid+tyrosine

(Building block formulae appear in Figure 1 or in the *Source Book*.)

For the glycosidic bond between monosaccharide building blocks use the β-1,4 link. (This is the one shown in Figure 1a. Do not worry much over this point, as glycosidic bonds really belong to Unit 3.) For the peptide bond use the α-COOH group, i.e. that attached to the α-carbon as shown in Figure 1b. (The other COOH group available in glutamic acid is never used in proteins but only in small peptides.)

SAQ 2 (*Objective 1*)   When a large number of building blocks condense together the result is not just a dimer—as in SAQ 1—but a macromolecule. Name the class of macromolecule formed in this way from molecules of the following:

(a) glucose

(b) d-ATP, d-GTP, d-TTP, d-CTP (d-ATP, etc., represent deoxyribonucleotides, i.e. those containing deoxyribose, not ribose)

(c) ATP, GTP, UTP, CTP

(d) aspartic acid, lysine, cysteine, etc.

SAQ 3 (*Objective 2*)   Would you expect the following macromolecules to have an informational or a structural/storage role? In (b) and (c), *n* denotes a large number.

(a)   See margin.

(b)   (Gly—Ser—Gly—Ala—Gly—Ala—)$_n$

(c)   (β-*N*-acetylglucosamine—)$_n$

SAQ 4 (*Objective 3*)   The bonding energy between two atoms, C and D, reaches a maximum at 20 kJ mol^{-1}. Does this suggest that the weak bonding involved is more likely to be hydrophobic, van der Waals or hydrogen bonding? (*Hint* Use the bond energies given in Table 1.)

SAQ 5 (*Objectives 5 and 6*)   Interaction between repressor† and operator† involves two classes of macromolecule, protein and nucleic acid. Assume for the moment that the amino acid histidine with a pK† of 6 lies in the repressor binding site on the DNA, and that this is the main residue involved when repressor binds to negatively charged regions of DNA. Will this binding be greatest at pH 5 or 7, assuming that the DNA is fully ionized at both pH values?

SAQ 6 (*Objective 5*)   How many potential opportunities are there for (a) H bonding (b) ionic bonding in the following macromolecular building blocks? Assume full ionization of any ionizable groups, even if this means rewriting the formulae as given here.

1   the amino acid serine

$$\overset{\text{OH}}{\underset{}{\overset{|}{\text{NH}_3^+ - \text{CH} - \text{COOH}}}}$$

2   a substituted monosaccharide

SAQ 3a

Gly—Ile—Val—Glu—Gln—Cys—Cys—Ala—Ser—Val—Cys—Ser—Leu—Tyr—Gln—Leu—Glu—Asn—Tyr—Cys—Asn

Phe—Val—Asn—Gln—His—Leu—Cys—Gly—Ser—His—Leu—Val—Glu—Ala—Leu—Tyr—Leu—Val—Cys—Gly—Glu—Arg—Gly—Phe—Phe—Tyr—Thr—Pro—Lys—Ala

3 a polypeptide chain containing a glutamic acid side chain

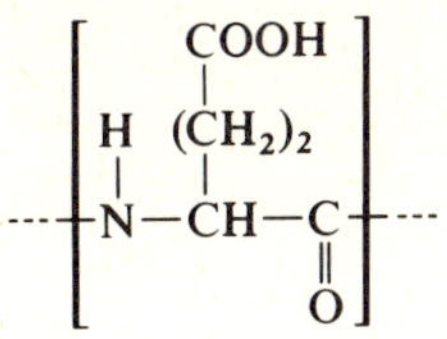

$$
\left[
\begin{array}{c}
\text{COOH} \\
\text{H} \quad (\text{CH}_2)_2 \\
\text{N}-\text{CH}-\text{C} \\
\parallel \\
\text{O}
\end{array}
\right]
$$

4 the nucleotide base adenine

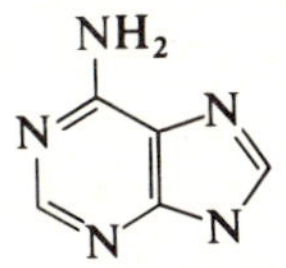

SAQ 7 (*Objective 4*) It is probable that the hormone† noradrenalin (Fig. 28a) exerts its effect on specific target organs by combining with receptors† found only on the membrane of cells that are sensitive to this hormone. Which *two* of the hypothetical noradrenalin analogues (b) to (e) in Figure 28 would you say showed most potential as noradrenalin-mimicking drugs?

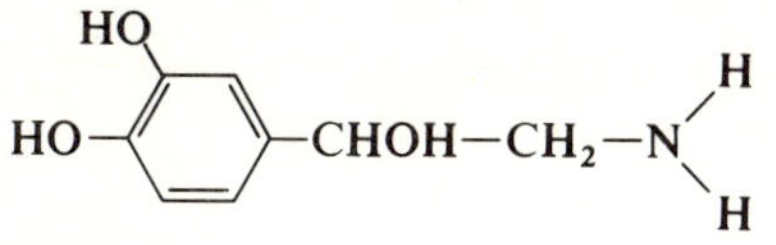

(a) Noradrenalin

(b)

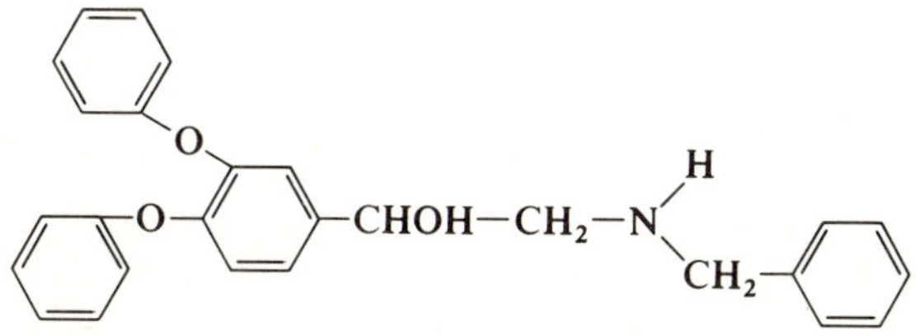

(c)

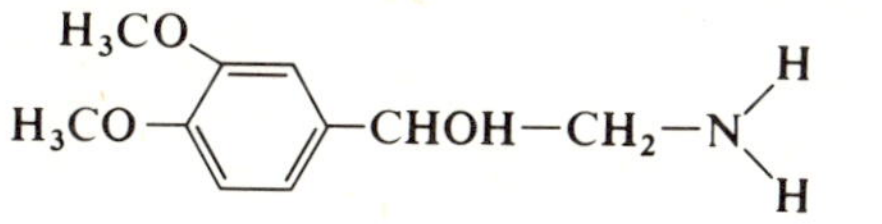

(d)

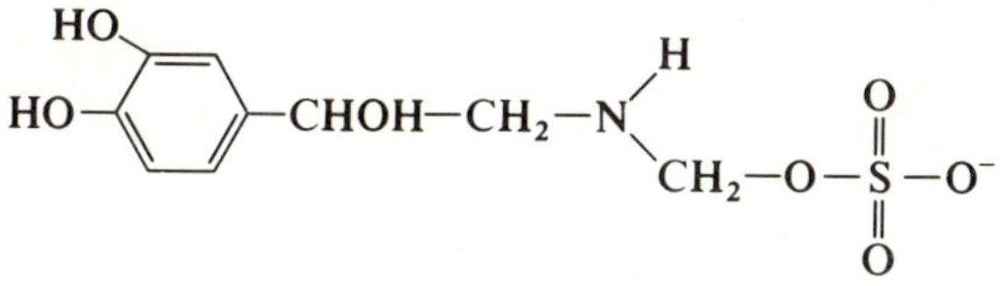

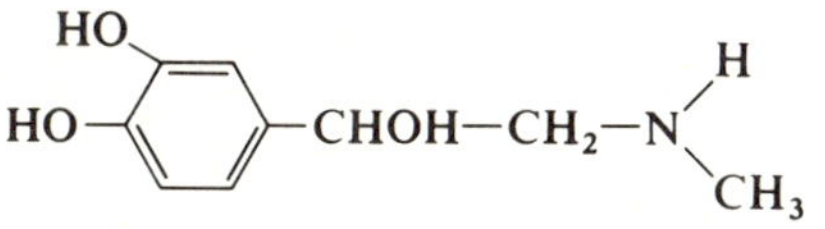

Figure 28 Noradrenalin, and noradrenalin-mimicking drugs.

(e)

SAQ 8 (*Objective 4*) A tripeptide hormone X occurs in brain extracts together with the enzyme Y responsible for its synthesis. Which of the following methods A or B is most suitable for obtaining a biologically active preparation of
(a) pure tripeptide X,
(b) pure enzyme Y?
Method A: gel chromatography of the brain extract at pH 7;
Method B: standing the brain extract overnight at pH 2, to allow protein Y to precipitate, and then centrifuging to separate protein (in precipitate) from tripeptide (in supernatant).

LIVERPOOL INSTITUTE OF
HIGHER EDUCATION
THE MARKLAND LIBRARY

SAQ 9 (*Objective 6*)   Imagine a simple viscometer where viscosity is measured by the time it takes a solution to flow between two points in a capillary tube. Which of the following solutions would have the slowest flow rate in such a viscometer:

(a) buffer alone,

(b) buffer containing dissolved ribonuclease,

(c) buffer containing 4 M urea and dissolved ribonuclease?

(Look up the urea reference in Table A2, if you have forgotten its effect on macromolecules.)

SAQ 10 (*Objectives 6 and 7*)   The single stranded DNA of phage ϕX174—a virus that attacks specific bacteria—is normally resistant to attack by bacterial exonucleases in the bacterial cell where it is found. (An exo-enzyme can hydro- lyse only from the free ends of a macromolecule, while an endo-enzyme can attack the middle of the backbone chain.)

(a) From your knowledge of the tertiary structure of DNA (Figure 18), which of the three forms does this suggest has been adopted by the phage DNA molecule? (Assume it is a closed circle without any nicks.)

(b) When endonuclease from an outside source is added to an extract of phage- infected bacteria, it snips the phage DNA open in the middle of the chain. This starts off rapid digestion by the bacterial exonuclease already present. Suggest a biophysical technique suitable for following the progress of this digestion.

SAQ 11 (*Objective 8*)   (This SAQ refers back to the Meselson and Stahl experi- ment described in Figures 19 and 20.) Would you expect the composition of DNA from heavy cells returned to a normal growth medium for just one division to be pure heavy ($^{15}N\,^{15}N$), hybrid ($^{14}N\,^{15}N$) or pure light ($^{14}N\,^{14}N$),

(a) if replication were conservative,

(b) if replication were semi-conservative?

SAQ 12 (*Objective 8*)   (This SAQ again refers to the Meselson and Stahl experi- ment.) Do the data in Figure 20 support conservative or semi-conservative replication?

SAQ 13 (*Objective 8*)   (This SAQ refers back to the data on PDC in Table 4.) Which of the eight lines of results in Table 4 can be taken in support of the following statements?

(a) Eb has a strong affinity for Ec.

(b) Ea, Eb and Ec form a spontaneously aggregating system.

(c) Ea and Ec have no complementary binding site for each other.

(d) Eb appears to form the central core on which the whole complex is built.

SAQ 14 (*Objective 7*)   Which of the techniques in the left-hand column would you choose for determining the molecular weights of the macromolecules in the right-hand column?

Technique	Macromolecule
polyacrylamide gel electrophoresis in SDS	(a) ribonuclease (this protein has no subunits)
sedimentation velocity (assume that either [η] or D is known if needed)	(b) haemoglobin
autoradiography	(c) the α subunit of haemoglobin
	(d) RNA from the 50S ribosomal 'subunit'
	(e) a large bacterial DNA

Pre-Unit test answers and comments

(Many of these answers can be supplemented by explanations in the *Source Book* or in previous Courses. References, when not cited here, can be found in Table A1.)

1 True. Elementary theories of bonding are described in S100, Unit 8, Section 8.4.1.

2 False. The statement reads true, however, if you interchange the terms *electrovalent (or ionic) bond* and *covalent bond*. You then have a definition of the key difference between these two bond types—in their pure form. As you will see in this Unit, however, *many covalent bonds are partially ionic*, owing to unequal sharing of electrons between the combining nuclei.

3 In order of decreasing polarity the compounds are:

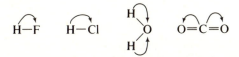

Arrows indicate the directions in which electrons (i.e. negative charge) will tend to predominate. Polarity is highest where there is a large difference in electronegativity between the combining atoms. For example, H—F $(4.0 - 2.1)$ is more polar than H—Cl $(3.0 - 2.1)$. (For polyatomic molecules like H_2O or CO_2 the net charge differences are admittedly harder to predict than for diatomic molecules like H—F and H—Cl, because molecular shape needs to be taken into account.)

4 (i) Deoxyribonucleotides, (ii) glucose, (iii) amino acids. (For further details, look up nucleic acid and polysaccharide in the *Source Book*.)

5   (i) $pH = -\log[H^+] = -\log 10^{-7} = 7$
    (ii) $pH = -\log 0.1 = -\log 10^{-1} = 1$
    (iii) $pH = -\log 0.01 = -\log 10^{-2} = 2$
Note that a *neutral* solution like pure water has a pH of 7 and that progressively more *acid* solutions have progressively *lower* pH values.

6 The dissociation reaction is:

$$CH_3COOH \rightleftharpoons CH_3COO^- + H^+$$

The equilibrium constant for any reaction is given by the ratio of product concentration to reactant concentration, i.e. [products]/[reactants], so that here:

$$K = \frac{[CH_3COO^-][H^+]}{[CH_3COOH]}$$

For further explanation see the *Source Book* under pK and equilibrium.

7 (i) True. You can deduce this from the *Source Book* entry under pK. Equation 1 shows that a strong (i.e. almost fully dissociated) acid will have a high concentration of free H^+ and A^-. Substitution into equation 2 then gives a high value for K.
(ii) True. Since p$K = -\log K$, then a strong acid (high K) will have low pK.

8 (i) True. At 50 per cent dissociation,

$$[\text{dissociated ion}] = [\text{undissociated ion}]$$

Substitution in equation 6 of the *Source Book* (under pK) then gives

$$pH = pK + \log 1 = pK + 0 = pK$$

(ii) Above the pK value. If the acid is mainly undissociated, the term log {[undissociated]/[dissociated]} will be large and positive. Substituting this into equation 6 gives pK greater than pH. Therefore at a pK value above the pH of the solution, the acid is largely undissociated.
(iii) Zero. If 50 per cent of the histidine is in the un-ionized form at pH 6, nearly 100 per cent will be in this form at pH 7.

9 CH_3—CH_3, $CHCl=CH_2$, $CH\equiv CH$. Rotation about covalent bonds becomes increasingly more difficult as single→double→triple.

10 True. These are examples of *optical isomers* (see *Source Book*).

11 (i) 340 nm. (See the definition of λ_{max} in the *Source Book* under chromophore.)
(ii) A. (See definition of molar absorbance in the *Source Book*.)

12 True. (See *Source Book*, under electrophoresis.)

13 True. The greater the enzyme *specificity*, the more precisely tailored must the substrate molecule be to fit the active site.

14 False. An allosteric inhibitor combines at a site *other than* the active site. However, if you replace the word 'allosteric' with the word 'competitive', the statement is a description of a *competitive inhibitor*.

15 No. The binding site of permease protein, just like the active site of enzyme protein, is highly specific. Usually a particular amino acid permease will reject not only sugar molecules, but all except a few closely related amino acids.

16 True. For further information, see under protein synthesis in the *Source Book*.

17 (i) DNA; (ii) mRNA; (iii) tRNA; (iv) DNA.

18 (i) Parallel. (See S2–1, Unit 1, upper Figure 20; and Figure 6 in this Unit.)
(ii) Perpendicular. (See Figure 9.27 in *Cell Structure and Function*, p. 205.)

ITQ answers

ITQ 1
(a) Substrate binds to enzyme.
(b) Antigen binds to antibody.
(c) Competitive inhibitor binds to enzyme.
(d) Passenger molecule binds to permease.
(e) Allosteric inhibitor binds to enzyme.
(f) Neurotransmitter binds to receptor protein.
(g) Hormone binds to receptor protein.

ITQ 2
(a) Protein and nucleic acid. (Although known as 'subunits', the 30S and 50S ribosomal components are something quite different from our definition of subunit.)
(b) Protein and lipid.
(c) Protein and nucleic acid.
(d) Protein and lipid.
In most of the multimeric complexes given here the subunits, whatever their macromolecular class, are specifically tailored to fit into one another. (With lipids, as you will see in Unit 3, this tailoring may be rather less precise than with the other macromolecules.)

ITQ 3 See Figure 7. (Note that you will find stearic acid as a stearoyl residue in both the fatty acyl branches of the phospholipid shown here.) The main candidates for H bonding are: $-C=O$, $-OH$, $-NH$, and $-NH_2$, and ring O or N atoms. (Figure 7 is on p. 20.)

ITQ 4 $-CH_3$ and $-CH_2-$ are both non-polar, whereas

$$-\overset{\frown}{C}=O \text{ and } -\overset{\frown}{O}-H$$

are polar. These last two contain atoms of unequal electronegativity (electron attracting power) and the shared electrons are polarized as shown.

ITQ 5 See Figure 7. The commonest non-polar residues responsible for van der Waals bonding in macromolecules are $-CH_3$, $-CH_2-$, and the aromatic† rings of amino acids and nucleotide bases.

ITQ 6 The value of n is only 2 rather than 7. Substitution into equation 1.2 shows that this will greatly reduce the dependence of F upon the interatomic distance. Because of this, the high dielectric constant of the environment (i.e. water) exerts its full effect. This is one reason why ionic bonds, which are potentially the strongest of the weak bonds, rarely contribute more than 4 kJ mol^{-1}.

ITQ 7

(a) $\quad -COOH \rightleftharpoons -COO^- + H^+$

(d) $\quad -NH_2 + H^+ \rightleftharpoons -NH_3^+$

(b) $\quad -O-\overset{\overset{O}{\|}}{\underset{\underset{O}{\|}}{S}}-OH \rightleftharpoons -O-\overset{\overset{O}{\|}}{\underset{\underset{O}{\|}}{S}}-O^- + H^+$

(e) $\quad HN{\bigtriangleup}N + H^+ \rightleftharpoons HN{\bigtriangleup}NH^+$

(c) $\quad -O-\overset{\overset{OH}{|}}{\underset{\underset{O}{\|}}{P}}-OH \rightleftharpoons -O-\overset{\overset{OH}{|}}{\underset{\underset{O}{\|}}{P}}-O^- + H^+$

The first three residues are acidic, and their dissociation into free H^+ plus a *negatively* charged residue follows the general acid dissociation equation shown in the *Source Book* under pK (equation 1). Basic groups can be visualized as acidic groups that have already dissociated, but in this case the result is free H^+ plus a neutral molecule. This neutral molecule is the basic group, and it has the ability to become *positively* charged by reversing the dissociation reaction, as shown here in (d) and (e).

ITQ 8 The sphere. The greatest R_G is given by the random coil, which is a highly flexible molecule with no unique 3-D structure. Examine Figure 10 again if you had difficulty in answering.

ITQ 9 Unfolding to a random coil increases the molecular excluded volume and hence the viscosity rises. (See Figure 10, and serum albumin in Table 3.)

ITQ 10 At about 38 °C. This is the point at which $[\eta]$ starts to rise sharply, indicating some change in conformation.

ITQ 11 A small molecule. This is not excluded from any part of the gel, and therefore takes the longer route and requires more eluting fluid to clear it from the column.

ITQ 12 The impurity is smaller, since it is nearer the meniscus and therefore moving more slowly.

ITQ 13 Superhelical, open-cyclic, linear. The linear form, being long, thin and flexible, will have the greatest MEV and hence, because of its resistance, the slowest sedimentation rate (and the smallest S value: 14S); conversely, the compact superhelical form will have the fastest sedimentation rate (and the largest S value: 20S). The intermediate open circle form has an S value of 16.

ITQ 14 Equation 1.9 suggests a choice between following the drop in concentration of one of the three reactants (pyruvate, CoA or NAD), or the corresponding rise in concentration of one of the three products (see previous Course[8]). In practice this second approach is nearly always more satisfactory.

ITQ 15 The colourless fraction is unlikely to contain any Eb or Ec if it has no lipoic acid or FAD. On the other hand, reaction 1.10 is what one would expect of an isolated Ea in which the natural acceptor (the lipoyl residue of step 2 in Figure 21) is replaced by an artificial acceptor such as ferricyanide. Therefore, the probable constitution is:

colourless 9S fraction = Ea
yellow 30S component = Eb+Ec

ITQ 16 The urea and pH treatments were kept as mild as possible (short exposures at low temperature) so that quaternary structure was disrupted without damage to secondary and tertiary structure.

ITQ 17

(a) Denaturing conditions stronger than the 4 M urea or the high pH mentioned in ITQ 16. We already know that Eb is still intact and enzymically active after these treatments. In practice, the detergent SDS (which you have met before in Section 1.5.3) or conditions of *low* pH (pH 2.8) both gave complete dissociation.
(b) Sedimentation equilibrium was used. Note, however, that the run had to be performed at pH 2.8, to prevent reassociation of the subunits. (Nowadays SDS gel electrophoresis might well be the method of choice.)

SAQ answers

SAQ 1 (a)

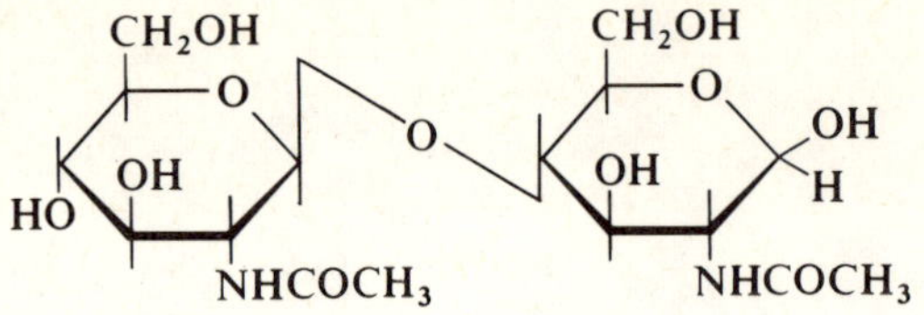

(b)

$$NH_2-\underset{\underset{H}{|}}{C}H-CONH-\underset{\underset{CH_3}{|}}{C}H-COO^-$$

(c)

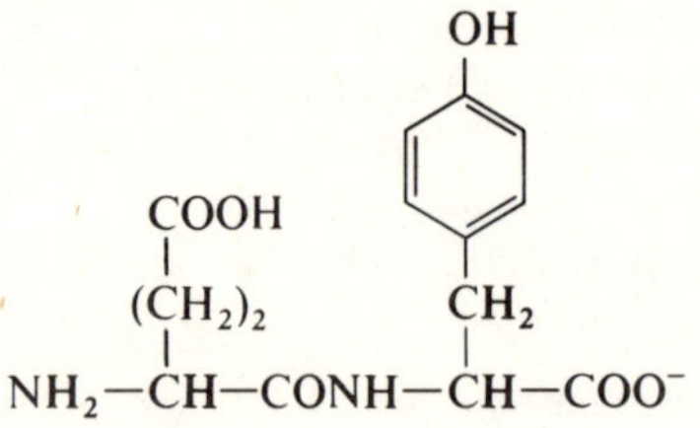

$$NH_2-\underset{\underset{(CH_2)_2}{\overset{COOH}{|}}}{C}H-CONH-\underset{\underset{CH_2-\langle OH \rangle}{|}}{C}H-COO^-$$

SAQ 2 (a) Polysaccharide (e.g. starch and glycogen which are both different forms of glucose polymer);
(b) DNA;
(c) RNA (note that UTP replaces TTP);
(d) a protein.

SAQ 3 (a) Informational. This is in fact *insulin*, a small globular protein hormone.
(b) Structural/storage role. This is in fact the main fibrous protein component of *silk*, and has very little variation in its primary structure compared with an informational protein like insulin.
(c) Structural/storage. This is in fact *chitin*, a structural polysaccharide of insect exoskeleton and fungi—see Unit 3. Note again the highly repetitive primary structure characteristic of structural/storage molecules.

SAQ 4 H bonding (see Table 1). When the combining atoms are correctly aligned, H-bond energy is far greater than van der Waals bond energy. (Hydrophobic bonding, as we said in Section 1.2.3, is nothing more than van der Waals bonding between C and D, strengthened by H bonding between expelled H_2O molecules.)

SAQ 5 Binding will be greatest at pH 5, where the histidine has its full positive charge (see answer to pre-Unit test question 8). At pH 7 histidine is un-ionized, with net charge zero.

SAQ 6

1 (a) Two: $-C{=}O$ (from COOH), OH

(b) Two: $-\underset{\underset{O}{\|}}{C}-O^-$ (from COOH), NH_3^+

2 (a) Five: $-C{=}O$ (twice), $-NH$, $O{=}\overset{|}{\underset{|}{S}}{=}O$ (twice), ring O

(b) One: $-O-\underset{\underset{O}{\|}}{\overset{\overset{O}{\|}}{S}}-O^-$ (see ITQ 7)

3 (a) Three: $-NH$, $C{=}O$ (twice)

(b) One: $-\underset{\underset{O}{\|}}{C}-O^-$

54

4 (a) Five: $\overset{\diagdown}{\underset{N}{\diagup}}$ (i.e. the four ring nitrogens), $-NH_2$

(b) None—unless NH_2 ionizes to $\overset{+}{N}H_3$, which in this particular molecule is very unlikely at biological pH values because of its high pK value.

SAQ 7 Compounds (c) and (e) show most potential as adrenergic drugs since they might still be able to fit into the specific hormone-binding site on the receptors. Compound (b) with its bulky aromatic rings, and compound (d) with its strongly charged sulphate groups are unlikely to fit in a pocket designed for noradrenalin.

SAQ 8 Method A for pure enzyme Y. Method B for pure tripeptide X. A tripeptide is not a macromolecule, has no weak bonding elements of structure and therefore cannot be denatured by extremes of pH. An enzyme like Y, being a protein macromolecule, is highly likely to be denatured at pH 2. Therefore method B is unsuitable for the preparation of *biologically active* Y.

SAQ 9 The buffer containing ribonuclease and urea will have the slowest flow rate. Urea, being a strong denaturant, will cause the ribonuclease to unfold from its compact native spherical form to a denatured random coil form of high viscosity.

SAQ 10 (a) Circular or superhelical. DNA in this form has no free ends and is therefore not attacked by exonuclease. (b) Viscometry (to measure $[\eta]$) or sedimentation velocity (to measure S value). Both these parameters are relatively simple to measure, and are sensitive to the changes in shape that follow as the partially-digested DNA unfolds.

SAQ 11 (a) Conservative replication would yield 50 per cent heavy DNA and 50 per cent light DNA. (b) Semi-conservative replication would yield—after just one division—only hybrid $^{14}N\,^{15}N$ DNA (see Figure 19).

SAQ 12 Semi-conservative replication. The buoyant density of DNA isolated from heavy cells which have been allowed just one division on a light growth medium, is shown in Figure 20b. As you can see, it falls midway between the positions of pure heavy and pure light DNA obtained from the calibration run (a), suggesting that its isotopic composition is also midway between the two, i.e. $^{14}N\,^{15}N$, as required by the semi-conservative model in Figure 19.

SAQ 13 (a) Line 7, by comparison with lines 1 and 3. Eb and Ec have recombined to give a single peak, with S value greater than that of either individual component.
(b) Line 4, by comparison with lines 1, 2 and 3. The single peak is again a 'new' one, with S value higher than those of the individual components.
(c) Line 8, by comparison with lines 1 and 3. No new peak appears, and the two peaks which are seen have S values characteristic of the original components.
(d) Line 4, by comparison with lines 8 and 5. Lines 4 and 8 together show that Eb is the missing link in the association of Ea and Ec. Line 5 is needed to confirm that the reconstituted complex resembles the original, and is not an artefact.

SAQ 14 (a) PAGE in SDS. This, like most empirical methods, is simpler to perform than sedimentation velocity runs, and we may assume sufficient standards are available.
(b) Sedimentation velocity. (PAGE is no good because the SDS would cause dissociation of the $\alpha_2\beta_2$ structure of haemoglobin, giving erroneously low results.)
(c) PAGE in SDS. Here dissociation is essential.
(d) PAGE in SDS (or sedimentation velocity, but see comment in (a)).
(e) Autoradiography. This is an exceedingly large molecule, so there would probably be insufficient standards of the right size range to use PAGE. Sedimentation velocity would be no good without a value for $[\eta]$—which is notoriously difficult to obtain for really large molecules.

Acknowledgements

Grateful acknowledgement is made to the following for material used in this Unit:

Figure 6b from M. D. Yudkin and R. Offord (1973) *Comprehensible Biochemistry*, Longman; *Figure 11 and Table 3* from K. E. Van Holde (1971) *Physical Biochemistry*, reproduced by permission of Prentice-Hall Inc., Englewood Cliffs, N.J., U.S.A.; *Figure 16* from R. J. Connett and N. Kirshner (1970) Purification and properties of bovine phenylethanolamine N-methyl transferase, *Journal of Biological Chemistry*, **245**, The American Society of Biological Chemists Inc.; *Figures 22 and 23* from M. Koike *et al.* (1963) α-Keto acid dehydrogenases: IV, *Journal of Biological Chemistry*, **238**, The American Society of Biological Chemists Inc.; *Figure 25* from F. Wold (1971) *Macromolecules: Structure and Function*, Prentice-Hall Inc.; *Figure 27* courtesy of Dr. John Cairns, Imperial Chemical Research Fund.

Unit 2 Conformation

Contents

Objectives

After completing this Unit you should be able to:

1 Explain in terms of bond strength and directionality, how hydrogen and hydrophobic bonding influence:
(a) stability of secondary structure in proteins and nucleic acids;
(b) their specificity of interaction.
(SAQs 1, 8, 9 and 10)

2 Pick out stretches of α-helix and β-pleated sheet in diagrams and models depicting tertiary structure in proteins.
(SAQs 4, 5 and 6)

3 Relate tertiary structure to
(a) stability and
(b) function in globular proteins and in tRNA.
(SAQs 2, 3 and 7)

4 Select from a given list, techniques suitable for studying conformation and conformational changes in macromolecules.
(SAQs 11, 12 and 13)

5 Draw conclusions about conformation from experimental data on spectroscopy, X-ray diffraction and chemical modification.
(SAQ 14)

Table A1

Assumed knowledge

Topic	Reference
absorption spectrum†	S24–,[5] Unit 3, Sections 3.3.1, 3.3.2
active site†	S2–1,[1] Unit 2, Section 2.3
allosteric enzyme	S2–1, Unit 5, Section 5.3; *C, S and F*,[2] p. 259. (Tested in S322,[3] Unit 1, pre-Unit test question 14.)
aromaticity†	S100,[4] Unit 10, Appendix 5
coenzyme (cofactor)†	S2–1, Unit 2, Section 2.4.1
delocalized electron	S100, Unit 10, Appendix 5
*electromagnetic radiation	S100, Unit 28, Section 28.2.1
energy level	S100, Unit 6, Section 6.5
enzyme–substrate complex†	S2–1, Unit 2, Section 2.4.1
enzymic catalysis	S2–1, Unit 2. (Tested in S322, Unit 1, pre-Unit test question 13)
*excited state	S100, Unit 6, Section 6.5
flexibility of macromolecules	S322, Unit 1, Section 1.3
frequency (i.e. velocity of light/wavelength)	see wavelength
haemoglobin	S100, Unit 18, Section 18.2.3
hydrogen bond†	S322, Unit 1, Section 1.2.2
hydrophobic bond†	S322, Unit 1, Section 1.2.3
hydrophobic residue†	S322, Unit 1, Section 1.2.3
induced fit theory of enzyme catalysis†	S2–1, Unit 2, Section 2.4.1
ionic bond†	S322, Unit 1, Section 1.2.4
molecular orbital	*Source Book*[6]
non-polar residue†	S322, Unit 1, Section 1.2.3
pH, p*K* and dissociation	*Source Book*; S2–1, Unit 2, Appendix 1. (Tested in S322, Unit 1, pre-Unit test questions 5–8)
polar residue†	S322, Unit 1, Section 1.2.3
*polarized light	S100, Unit 10, Appendix 2; Unit 28, Section 28.6
primary structure	S322, Unit 1, Section 1.1.3
protein synthesis†	S100, Unit 17, Sections 17.4, 17.9; S2–1, Unit 6, Section 6.3. (Tested in S322, Unit 1, pre-Unit test question 17)
quanta (quantum)	S100, Unit 29, Section 29.2.1
quantum number	S100, Unit 6, Section 6.5
quaternary structure	S322, Unit 1, Section 1.1.3
replication†	S100, Unit 17, Section 17.2.2
secondary structure	S322, Unit 1, Section 1.1.3
specificity of macromolecular interaction	S322, Unit 1, Section 1.3.1
tertiary structure	S322, Unit 1, Section 1.1.3
tRNA	S2–1, Unit 6, Section 6.3; *C, S and F*, pp. 383–385
van der Waals bond†	S322, Unit 1, Section 1.2.3
*vector (electric vector)	S100, Unit 28, Section 28.6
wavelength	S100, Unit 2, Section 2.3.1
wave nature of light	S100, Unit 29, Section 29.1

* Items marked with an asterisk (*) are required only for the optional reading in Appendices 1 and 2.

† These items are also to be found in the *Source Book*.

[1] The Open University (1972) S2–1 *Biochemistry*, The Open University Press.

[2] Loewy, A. G. and Siekevitz, P. (1969) *Cell Structure and Function*, 2nd edn, Holt, Rinehart and Winston.

[3] The Open University (1977) S322 *Biochemistry and Molecular Biology*, The Open University Press.

[4] The Open University (1971) S100 *Science: A Foundation Course*, The Open University Press.

[5] The Open University (1973) S24—*An Introduction to the Chemistry of Carbon Compounds*, The Open University Press.

[6] The Open University (1977) S322 *Source Book for Biochemistry and Molecular Biology*, The Open University Press. Forms part of the supplementary material for this Course.

Table A2

List of scientific terms and concepts in Unit 2

Study guide for Unit 2

This Unit falls into two parts. The *principles* of conformation (Part I) form the most important Section and you should direct a major part of your energies here. The *experimental techniques* (Part II) are essential in that without them none of the first part could ever have been written, but if pressed for time you may cut down here. You should, however, try to appreciate the scope of each technique and see how each one, particularly X-ray crystallography, has contributed to our knowledge of conformation. Table 5 should be useful as a summary of Part II.

Although there may seem to be more emphasis on proteins than nucleic acids, this Unit provides important background information on nucleic acids for Unit 11. Therefore the Sections on nucleic acids should receive the same attention as those on proteins, except where possible omission items are indicated in the Study comments.

The model building kit is essential for a full understanding of conformational angles (Section 2.1). The stereoviewer is needed for Sections 2.1, 2.2 and 2.8, and for some of the SAQs related to these Sections (Objective 2). The slides needed for each Section are listed in the Study comments. The *Source Book** is not referred to in Section 2.1, but thereafter you may need it at various places throughout the Unit for formulae of amino acids and nucleotides. Items to be found in the index to the *Soure Book* are marked with a dagger (†). The television programme on lysozyme structure is closely integrated with this Unit, particularly in Sections 2.2 (tertiary protein structure) and 2.8 (X-ray diffraction).

Assumed knowledge is similar to that for Unit 1, and the same pre-Unit test questions apply. Those particularly relevant to Unit 2 are indicated in Table A1 of this Unit, where references to previous Courses, Unit 1 of this Course or to the *Source Book* are given. References to this Table in the text are indicated by a superscript A.

---

* The Open University (1977) S322 *Source Book for Biochemistry and Molecular Biology*, The Open University Press.

Part I   Principles of conformation

Introduction to Part I

We start the Unit by summarizing what is known about conformation in both proteins and nucleic acids. This is essentially the same thing as describing secondary and tertiary structure. The points to look out for are the regular features (secondary structure) common to all macromolecules of that class, and the irregular features unique to each particular protein or nucleic acid. In fibrous proteins and DNA, the regular sections tend to predominate. In other macromolecules like tRNA and the globular proteins, the regular features are generously interspersed with irregular sections, the whole folding up to a tertiary structure with infinite possibilities for variations in shape. This may reflect the function of these molecules, which require rather more specificity in their interaction with other molecules than do, for example, the fibrous proteins which are mainly structural. You should be able to give examples of this difference in complexity between macromolecules of different function, and you should also note another distinguishing characteristic—uniqueness of conformation.

In theory, a protein can fold into any one of a vast number of conformations, but in practice these are limited to the single form which has maximum stability for a particular set of conditions. This unique conformation is predetermined, as we shall see, by the sequence of amino acids in the primary structure. tRNA also has a unique conformation, determined here by the sequence of nucleotide bases. However, most other nucleic acids are able to assume more than one stable conformation, depending partly on their association with other macromolecules such as protein. This will become apparent in the discussion of higher order structure in DNA.

Within narrow limits, even proteins and tRNA can undergo small changes in conformation, caused by rotation about the many single bonds in the molecule; this is the flexibility described in Unit 1, Section 1.3.2, to which you may now like to refer. These conformational changes are triggered by any change in environmental conditions. In the cell such changes are small, induced by binding of ligand* or by small changes in pH or temperature. They may be confined to areas around the ligand binding site, or may be relayed (as in allosteric proteins) to a more remote part of the molecule. In either case, the conformational changes are very different from the disturbances, verging on denaturation, which can be brought about by larger changes in pH or temperature. Unphysiological though these latter conformational changes may be, they have provided much information on macromolecular structure. In particular, we shall be describing the thermal denaturation of DNA and its role in studying nucleic acid structure.

2.1   Secondary structure in proteins

> **Study comment** This is one of the most important Sections of the Unit and should not be omitted. You will need Stereoslides 3, 4 and 5, and the model building kit. The kit should help you to understand conformational angles, and the slides are essential for fulfilling Objective 2.

Proteins in their native** state are never fully extended, but are folded in a precisely defined fashion to form either a *fibrous protein* such as silk or keratin, or an approximately spherical *globular protein* like haemoglobin and many of the enzymes. Both classes of protein have regular folding patterns which are stabilized by H bonding: the α-helix and the β-pleated sheet (often abbreviated to

fibrous protein
globular protein

* A *ligand* may be defined as any small molecule which binds specifically to a macromolecule. Common examples of protein ligands include substrates, inhibitors and coenzymes.

** The *native* state of a macromolecule may be defined as the biologically active conformation (see also Section 2.2.1).

α- and β-structures). These can be seen most clearly in the fibrous proteins which, as we explained in Section 1.1.2 of Unit 1, have a mainly structural role and therefore a highly repetitive folding pattern. Most globular proteins, on the other hand, have only short stretches of α- and β-structure, interspersed with irregular folding patterns.

All regular secondary structures are stabilized by H bonding, and you should be aware throughout this Section of the consequences of the marked directionality of this bond. Weak H bonds, where the combining atoms are not strictly aligned, do not appear in α- or β-structures.

2.1.1 Conformational angles*

The most basic fact about protein structure is one that not everyone is immediately aware of—that the peptide bond is planar. Look at the tripeptide in Figure 1, focusing in particular on the amide bond, $-CONH-$, which may also

be written as

$$-\overset{\overset{\displaystyle O}{\displaystyle \|}}{C}-\underset{\underset{\displaystyle H}{\displaystyle |}}{N}-$$

(a)

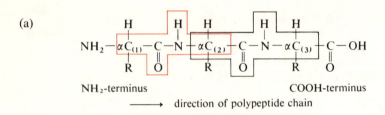

NH₂-terminus   COOH-terminus

⟶ direction of polypeptide chain

(b)

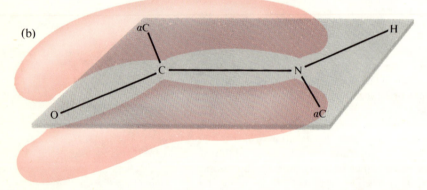

Figure 1 Planarity of the peptide bond. (a) Structural formula of a dipeptide, outlining the six atoms in the plane of each peptide bond. (b) Detail of one peptide bond, showing molecular orbital (red shaded area) encompassing C, O and N atoms.

It is the nature of the link between the carbonyl C and the N that is important. Because of resonance† it is not, as you might think, a pure single bond, but has a partial double bond character. This is because the lone pair electrons from the $C{=}O$ group are not confined to the area between these two atoms but are distributed in a molecular orbital[A]** extending over all three atoms, O, C and N. This orbital is shown in Figure 1b, and the conformational angles ϕ and ψ in Figure 49 (p. 58).

> **ITQ 1** Consider the tripeptide A—B—C shown in Figure 2. The planes of the three peptide bonds are outlined and the atoms which belong to each particular plane are subscripted A, B or C. Given the lack of flexibility between atoms within any one plane, about which of the following bonds would you expect there to be free rotation: $N_A{-}C_A$, $C_A{-}\alpha C_B$, $\alpha C_B{-}N_B$, $\alpha C_C{-}N_C$, $N_B{-}C_B$? (The answer appears on p. 58).

* Conformational angles are further discussed in the TV programme on polysaccharides.

** Further information on terms marked with a superscript A is to be found where listed in Table A1.

† Terms marked with a dagger (†) are to be found in the index to the *Source Book*.

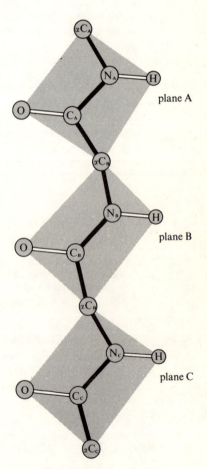

plane A

plane B

plane C

Figure 2 Data for ITQ 1, on the relative orientation of adjacent peptide bonds in a tripeptide.

MODEL BUILDING EXERCISE 1 The conformational angles ϕ (phi) and ψ (psi) are very much easier to visualize from a three-dimensional model. Make the dipeptide shown in Figure 3a.* This is the central part of the dipeptide of Figure 1a including only the 11 atoms within the 'box', plus the R groups on αC-1 and αC-2.** Ideally, the middle-length straws should be used for the C—N and C—O bonds, to simulate their partial double bond character. This point is not essential, but do not forget to use *trivalent* C for the carbonyl C, and *tetravalent* C for the α-carbon. We suggest you use the yellow sulphur atom to represent R groups. *Check* Do the six atoms in the plane of each peptide bond

$$\begin{array}{c} \alpha C \\ \diagdown \\ H \diagup N - C \diagup \alpha C \\ \diagup \quad \diagdown O \end{array}$$

really lie in the same plane? This is a very important point. (Note that the second αC is common to both planes.)

Now arrange the model as in Figure 3a, where both conformational angles are zero. For convenience, hold the model by its central α-carbon, approximately parallel to a table surface.

Check Is the yellow R group on αC-1 projecting approximately downwards?

To illustrate conformational angle ϕ, rotate the nearer peptide bond plane anticlockwise through 180°, to give the model shown in Figure 3b. The dipeptide now has the conformation $\phi = 180°$, $\psi = 0°$. To illustrate the other angle ψ, rotate the second (more distant) plane clockwise through 60°, to give the model shown in Figure 3c.‡ This has conformation $\psi = 180°$, $\psi = 60°$.

To check that you have mastered the conformational angles, return the dipeptide to its starting position ($\phi = \psi = 0°$, as in Figure 3a) and try ITQ 2.

ITQ 2 One of the two conformations, $\phi = 40°$, $\psi = 315°$ and $\phi = 130°$, $\psi = 120°$, describes an α-helix and the other a β-pleated sheet. Use the dipeptide model from model building exercise 1 to make each of these conformations. Then decide which description fits which of the conformations you have just made.

Hint Remember that in a β-sheet the dipeptide would form part of a ribbon, and in an α-helix it would be part of a spiral staircase-like structure (see Figure 5).

2.1.2 Helical parameters

It is because the two conformational angles can in theory vary all the way from 0 to 360° that there is such a tremendous range of protein conformations. If the values for ψ (and for ϕ) are the same at each α-carbon, the polypeptide chain twists automatically into a helix. The α-helix (where both angles approximate to 125°) is one such structure commonly found in proteins. It makes up the bulk of fibrous proteins such as wool and keratin, and can be found, at least in short stretches, in most globular proteins.

Although any helix where the repeat units are of equal length can be fully defined by the values of ψ and ϕ, it is easier to visualize the structure in terms of helix *pitch*. This takes into account both the number of repeat units per turn, n, and

helix pitch

---

* Figure 3 has also been printed in the *Notes on the Use of the Model Building Kit*. Open University students may find it more convenient to work from this, rather than keep turning to p. 10 where it is printed in this Unit. Figure 3 is also used for model building exercise 2, and again Open University students will probably prefer to have the Figure in the *Notes on the Use of the Model Building Kit* in front of them rather than having to turn back to p. 10 every time they need to consult the Figure.

** αC denotes α-carbon. C-1 denotes the carbon atom numbered 1, C-2 the carbon atom numbered 2, etc.

‡ These angles can be estimated accurately enough if you remember that 90° is one-quarter of a full turn, 180° is half a turn, and 270° is three-quarters.

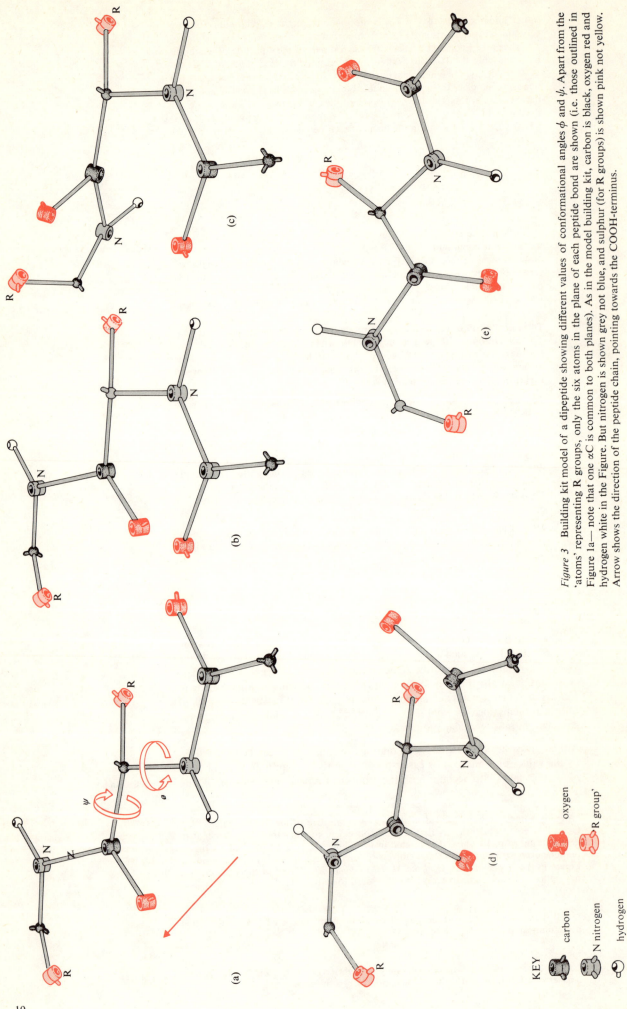

Figure 3 Building kit model of a dipeptide showing different values of conformational angles ϕ and ψ. Apart from the 'atoms' representing **R** groups, only the six atoms in the plane of each peptide bond are shown (i.e. those outlined in Figure 1a— note that one αC is common to both planes). As in the model building kit, carbon is black, oxygen red and hydrogen white in the Figure. But nitrogen is shown grey not blue, and sulphur (for **R** groups) is shown pink not yellow. Arrow shows the direction of the peptide chain, pointing towards the COOH-terminus.

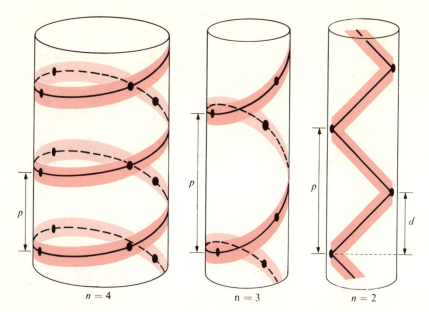

Figure 4 Helices of different pitch; *d* is the vertical distance between residues (shown as black circles), *n* the number of residues per turn of helix, and *p* ($n \times d$) is the pitch.

the vertical distance travelled by each unit, *d*; the pitch, *p*, is then the product of these two. Figure 4 should make this clear. Here we show three helices where *n* is 2, 3 and 4. The α-helix is a compromise between these last two, with an *n* value of 3.6. The β-pleated sheet has an *n* value of 2, which flattens the staircase-like structure to a zigzag ribbon.

What is it that makes the values 2 and 3.6 particularly popular amongst proteins? The answer is that this permits the maximum number of strong stabilizing hydrogen bonds to be formed. Remember from Unit 1 that H bonds, unlike other weak bonds, are highly directional, i.e. maximum binding energy results when the two electronegative atoms on either side of the hydrogen lie in the same straight line. In the α-helix an *n* value of 3.6 produces just this situation by bringing the C=O of one peptide bond to lie directly in line with the N—H, 3.6 residues below. This alignment occurs easily if there is one turn of helix every 3.6 residues. The helix is then stabilized by H bonds stretching up and down the length of the 'stair-well', as you can see in Figure 5. In contrast, the β-pleated sheet (Figure 6) is stabilized by H bonds at right angles to the polpypeptide chain. It is now time to look at these secondary structures in more detail.

2.1.3 β-Pleated sheet

The ribbon may be considered as the basic unit of a β-structure. This has two polypeptide chains in the β conformation, usually lying antiparallel, i.e. one starting at the NH_2-terminus and the other at the COOH-terminus. Globular proteins tend to have narrow stretches of β-structure, often only two chains across, while fibrous proteins like silk have extensive sheets formed of many ribbons lying side-by-side. In the complete fibre many such sheets are stacked above one another (see Figure 7). Figure 6 shows the details of individual chains, while Stereoslide 3 shows an extensive sheet of β-structure.

NOW LOOK AT STEREOSLIDE 3 IN DETAIL.

Only the N, αC and C=O atoms are shown. These are not individually colour-coded, as in your model building kit, but are all blue spheres. The αC can be recognized by the two 'spare' bonds projecting from it. The smaller of these represents the αC—H bond and the larger the αC—R bond. The C=O group can be recognized by its H bond (in red) projecting from the O. One of the poly-peptide chains in this slide has a short section printed in orange. This is to indicate the extent of a single amino acid residue,

$$-N-\alpha C-C-$$
$$\overset{\|}{O}$$

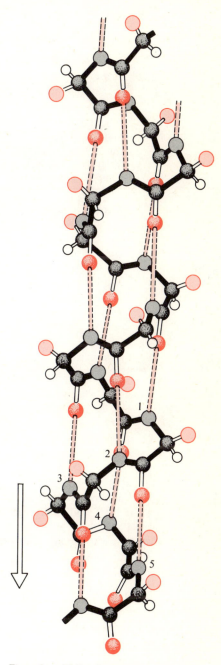

Figure 5 α-Helix. Arrow shows direction of polypeptide chain (heavy black line), pointing to the COOH-terminus. Numbers (1–5) indicate N atoms of adjacent amino acid residues in a short section of the chain. Note that H bonds run parallel to the helix axis, e.g. from O of residue 1 to N of residue 5. Atoms are presented as follows: oxygen, red; nitrogen, grey; carbon, black; hydrogen, white; R groups, pink. Hydrogen bonds are shaded pink. Hydrogen atoms attached to N have been omitted.

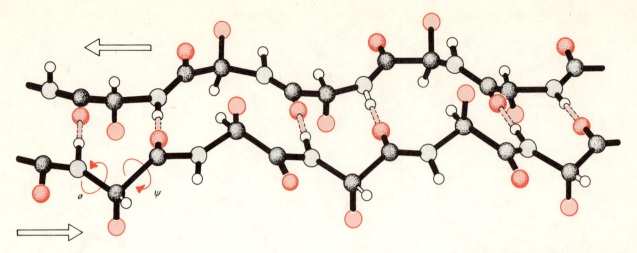

QUESTION Do the chains lie head-to-tail (one starting at the NH_2-terminus and the other at the COOH-terminus) or head-to-head (all starting at the NH_2-terminus)? Note that a chain starting at the NH_2-terminus reads $-NH-\alpha CHR-CO-$; a chain starting at the COOH-terminus reads $-CO-\alpha CHR-NH-$.

ANSWER Head-to-head.* All the chains run upwards from the bottom of the slide. For example, the last orange group in the second nearest chain is $-C=O$, and the chain then reads $-\alpha CHR-NH-CO-$ (this second O being hidden in this view).

Figure 6 β-Pleated sheet. Pair of anti-parallel polypeptide chains (arrows indicate chain direction). Conformational angles ϕ (40°) and ψ (315°) are indicated. Atoms and H bonds are colour coded as in Figure 5.

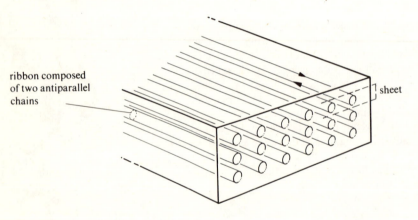

ribbon composed of two antiparallel chains

sheet

Figure 7 Diagram to show position of individual polypeptide chains in a fibre of β-pleated sheet structure.

Two kinds of weak bonding hold together the polypeptide chains in a β-structure. There is H bonding between the two chains in a ribbon pair, and van der Waals bonding between adjacent sheets stacked vertically in the fibre. The directionality of this weak bonding is best seen with a model.

MODEL BUILDING EXERCISE 2 The β-pleated sheet. For this exercise you will need to start with the same dipeptide as before (model building exercise 1), but with the second and third α-carbons fully substituted with H and R groups. (Use the yellow 'S atom' again for R.) Begin with the model in the conformation $\phi = \psi = 0$, holding it by the second α-carbon so that it lies approximately parallel to a table surface.

Check (i) Are there six atoms in the plane of each peptide bond

$$\begin{pmatrix} \alpha C & & \alpha C \\ & N-C & \\ H & & O \end{pmatrix}?$$

(ii) Does the yellow R group point slightly downwards on the second (central) α-carbon, and upwards on the third α-carbon?

Now form the β conformation ($\phi = 40°$, $\psi = 315°$), by rotating the nearer plane clockwise through 40° and the further plane anticlockwise through 315°. This gives the model shown in Figure 3d. Since 315° is almost a full 360° revolution,

---

* This head-to-head arrangement is known as *parallel*, while the alternative head-to-tail is *antiparallel*.

you will find that neither plane has tilted very much from its starting position. The sideways view (Figure 3e) shows a slightly V-shaped molecule. (This is the V that gives the 'pleated' look, clearly visible in Stereoslide 3.)

NOW TRY ITQS 3, 4 AND 5, WHICH CONCERN THE WEAK-BONDING POSSIBILITIES OF THE DIPEPTIDE IN THIS MODEL.

> **ITQ 3** In a sheet of polypeptide chains all in the β conformation, adjacent chains are held together by H bonding. How many H bonds could be formed on either side of the dipeptide you have just made in model building exercise 2?

> **ITQ 4** Imagine a second dipeptide, also in the β conformation, placed alongside the first to form a ribbon pair. Will the H bonds between these dipeptides be strong or weak? (Give a reason for your answer.)

> **ITQ 5** Imagine that the second dipeptide is placed on *top* of the first, as in a pile of stacked β-sheets. What groups are available for weak bonding between these dipeptides?

2.1.4 α-Helix

An α-helix cannot be so easily illustrated with the model building kit as can the β-sheet, so we have taken stereoslides of some more robust models.

NOW STUDY STEREOSLIDE 4.

Here you see the same ball-and-spoke model as in Stereoslide 3, but painted white. The N, αC and C=O atoms are white spheres, and as before the two spare αC bonds are to H and R. The orange section represents the extent of one residue.

> QUESTION   In this slide does the polypeptide chain run upwards, starting from the NH_2-terminus, or downwards?

> ANSWER   Upwards. For example, the first white atom to the right of the orange segment is N (note its downward pointing H bond) followed by αC—CO—.

What is particularly clear from this slide is the way the R groups—to be imagined at the end of the larger αC bond—project out from the main chain. When we come to tertiary structure we shall be describing how this helps bind the α-helix to adjacent stretches of the polypeptide chain. Note also how the H bonds (in red) run virtually parallel to the helix axis.

NOW STUDY STEREOSLIDE 5.

This slide shows what happens when these stabilizing H bonds are 'removed': the helix tends to fall apart. (The precarious balancing of the 'de-hydrogen bonded' section may not come across in the slide, but certainly we dared not remove any more for fear of collapsing the whole model.) The polypeptide chain again runs upwards from the bottom of the slide, and here the colour coding is the same as in your building kit model—nitrogen atoms are painted blue, carbon black and oxygen red. This should make it easier to see that there are just under four residues (3.6 to be exact) per turn of helix. (This is the helical parameter n of Figure 4.)

> QUESTION   Do the H bonds from nitrogen project backwards towards the NH_2-terminus at the bottom of the slide or forwards to the COOH-terminus?

> ANSWER   Backwards to the NH_2-terminus.

Conversely, the H bonds from O project forwards, and with this model you should be able to make out that there are three intervening residues before the N from the fourth residue, with which the O forms its H bond. Take the uppermost H-bonded N (identifiable because its H bond has been slightly skewed under the strain of 'de-bonding'!) Let this be the first atom of, say, residue 5.

Moving backwards down the chain you should be able to see that there are three N atoms (residues 4, 3 and 2) before the O component of the skewed H bond is reached on residue 1. (H bonding between C=O of residue 5 and N—H of residue 1 can be more clearly seen in Figure 5.)

2.1.5 Proline, the α-helix breaker

All the protein building blocks except one are capable of fitting into an α-helix. The exception is proline, the so called α-helix breaker. This is an imino rather than an amino acid* and, as you can see in Figure 8, it has no H on the nitrogen

proline

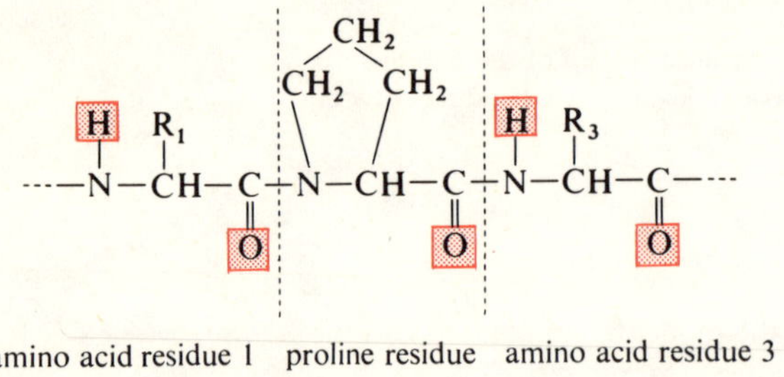

amino acid residue 1   proline residue   amino acid residue 3

Figure 8 Proline-containing tripeptide fragment of a polypeptide chain. Red shading indicates atoms that would, in an α-helix, be involved in H bonding. Note absence of H on the N of the proline peptide bond.

of the peptide bond. It therefore cannot contribute to an H-bond stabilized structure in the same way as the amino acids.

NOW YOU COULD TRY SAQ 1, ON P. 57.

2.2 Tertiary structure of proteins

Study comment You will need Stereoslides 1, 2, 6 and 7 here, and the *Source Book*'s table of amino acid formulae will be required for the ITQs. This Section is the one that is mostly closely linked with the TV programme on lysozyme structure. It is concerned mainly with Objectives 2 and 3.

The fibrous proteins with their low information content are often structural macromolecules, requiring mechanical strength. They are composed almost entirely of regular H-bonded features like the α-helix and β-sheet. In this Section we are concerned only with globular proteins, where the folding pattern is far more complex, reflecting the need for a more intricately designed 3-D shape. Their tertiary structure includes stretches of both α- and β-structures, interspersed with irregular sections which cannot be described in terms of any repetitive H-bonded features.

The final folding pattern of any globular protein must satisfy two criteria, *stability* and *maximum efficiency of function*, and we shall discuss each of these in turn.

2.2.1 Folding for stability

The drive to form internal hydrophobic bonds

All the proteins so far analysed in any detail tend to fold up according to one general rule: *maximum internal hydrophobic bonding gives minimum free energy.***

This means that in a polar environment like water a protein will tend to behave like an oil drop, folding up so that its non-polar residues^ are inside and its

---

* Despite this, proline appears in the *Source Book* with the amino acids.

** This rule is based on the interpretations of high resolution X-ray data (see Section 2.8) from water-soluble enzymes. Not all proteins, of course, are found in an aqueous environment. In the predominantly lipid surroundings of a membrane, for example, one might expect quite a different stability rule, but so far our knowledge of these other situations is not very far advanced.

14

polar residues outside. Although biochemists may have been oversimplifying in describing the cytoplasm as a simple aqueous solution (see Unit 7) most of the water-soluble globular proteins so far studied do approximate to the oil drop analogy. The reason for this is thought to be as follows. Hydrogen bonds, because of their strong preference for certain orientations, exert a directing influence over protein structure, but hydrophobic residues[A] are responsible for most of the bond energy. By congregating in the interior of the molecule they not only promote van der Waals bonding amongst themselves, they also expel polar residues and free water molecules to the exterior, where these in turn have maximum opportunity for bonding amongst themselves. The result is that hydrophilic (i.e. polar) amino acid residues tend to be found on the outside of the molecule, and hydrophobic inside. The TV programe on lysozyme structure illustrates this point.

internal hydrophobic bonding

Disulphide bonds

Although weak bonding is the driving force behind protein folding, covalent bonding may play some part in holding together the tertiary structure once it is formed. This is true of proteins containing cysteine, the amino acid with the free —SH group. Where the folding pattern brings two such groups to lie opposite one another in the tertiary structure, a disulphide bond can easily form between them, as in Figure 9. This is a covalent not a weak bond, and proteins held together by both weak bonding and disulphide bridges tend to be rather stable. They cannot, for example, be denatured without first reducing the —S—S— bridge back to two free —SH groups.

disulphide bond

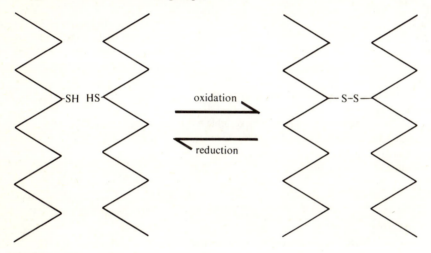

Figure 9 Disulphide bond formation.

Native versus denatured

When a protein folds under the driving force of hydrophobic bonding it assumes a unique, biologically active conformation. This is the *native* state of the protein, held together by weak (and sometimes also disulphide) bonding. Without these stabilizing bonds the protein falls apart to its *denatured* state. Tertiary structure is lost, leaving an extended polypeptide chain of no fixed conformation. This is the random coil described in Unit 1.

native state

A common protein denaturant is urea, a small molecule of formula

$$\begin{matrix} NH_2 \\ \\ NH_2 \end{matrix} \!\!\! > C = O,$$

thought to act by disrupting H bonding in the native protein and its surrounding layer of water.

Denaturation/renaturation

So strong is the driving force for internal hydrophobic bonding that even a fully denatured polypeptide chain can be made to fold spontaneously back into its native conformation. This observation is what lies behind one of the central dogmas of protein structure: *primary structure dictates tertiary structure*, i.e. all the information for folding a protein chain into its final shape is carried in the primary structure. This would allow a growing polypeptide chain to fold spontaneously as it came off the ribosome assembly line—an attractive but probably much oversimplified picture of events *in vivo*. *In vitro*, there is now substantial evidence in support of the dogma from denaturation/renaturation experiments with many proteins.* Addition of a denaturant like urea disrupts the tertiary structure of a native protein, leaving a flexible, biologically inactive polypeptide chain of no fixed conformation. If the dogma is correct, removal of denaturant should restore activity. This assumes, of course, that the extended polypeptide chain carries enough information to refold to its original conformation. Results, as you will see in Unit 4, show that very careful removal of denaturant can often restore almost 100 per cent activity. This suggests that primary structure does indeed dictate tertiary structure. A denaturation/renaturation cycle is shown diagrammatically in Figure 10.

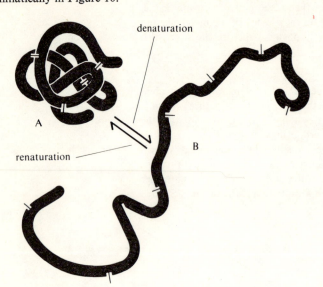

Figure 10 Denaturation and renaturation of a protein. (A) Native conformation; (B) extended random coil conformation.

Under the right conditions, nearly 100 per cent of the protein can spontaneously re-form the active conformation. We may therefore further assume that this unique conformation is one of the stablest of the enormous number available to any polypeptide chain. If the chain starts to fold as soon as it is partly formed, the final conformation may not necessarily be the stablest in absolute terms, but the stablest that can form in a finite time.

NOW YOU COULD TRY SAQS 2 AND 3.

A specific example: triosephosphate isomerase

In Section 2.1 we described how fibrous proteins are held together by interlocking weak bonds between the chains. The principle is exactly the same in globular proteins, where all residues come from the same polypeptide chain. We can show this with a specific example, the enzyme triosephosphate isomerase (TIM). This controls the interconversion of two intermediates of the glycolytic pathway, dihydroxyacetone phosphate and glyceraldehyde-3-phosphate (see Table 1). The enzyme approximates to a sphere, within which the main chain winds back and forth across the molecule. Nearly 80 per cent of the residues are arranged in regular secondary structure. There are eight β-pleated sheets, forming a solid hydrophobic core of intramolecular β-structure. As you can see in Figure 11, these sheets are joined at the ends by lengths of α-helix, which form an outer cylinder round the molecule. A helix, like the β-sheet, projects its R groups out into the environment, but any helix in this outer cylinder of TIM will lie between two quite different environments—water on the outside and hydrophobic core on the inside. This should influence the arrangement of its sidechains (the R groups), as you can see in ITQ 6.

ITQ 6 Take a helix on the edge of an imaginary globular protein which, like TIM, has a hydrophobic core. If the residues jutting out from one side of the helix are the amino acids alanine, glycine, phenylalanine and tryptophan,** and those jutting out from the other side are glutamic acid, lysine and aspartic acid, which side would you expect to pack next to the hydrophobic core?

* The first and classic renaturation experiment was done on ribonuclease (see Unit 4).

** Amino acid formulae can be found in the *Source Book*.

16

TABLE 1 Reactions catalysed by the enzymes described in Section 2.2

Enzyme	Reaction catalysed*

TIM

$$\begin{array}{c}\text{CHO}\\|\\\text{CHOH}\\|\\\text{CH}_2\text{OPO}_3^{2-}\end{array} \rightleftharpoons \begin{array}{c}\text{CHOH}\\|\\\text{C=O}\\|\\\text{CH}_2\text{OPO}_3^{2-}\end{array}$$

glyceraldehyde-3-phosphate   dihydroxyacetone phosphate

lysozyme

D ring   E ring

part of an *N*-acetyl glucosamine–*N*-acetyl muramic acid (NAG–NAM) chain

carboxypeptidase

$$\begin{array}{cc}\text{R} & \text{R}\\ | & |\\ \text{-----NHCHCO} & \text{NHCHCOO}^-\end{array}$$

COOH-terminal
end of polypeptide chain

* Arrows denote site of cleavage; dotted lines denote part of a long chain. For carboxypeptidase, R denotes amino acid sidechain. For lysozyme, R denotes $-\text{CH}-\text{COOH}$.
                CH_3

Although the distribution of polar versus non-polar residues is seldom as clear-cut as this in any real protein, ITQ 6 does illustrate the kind of bonding that joins together adjacent stretches of polypeptide chain in the tertiary structure.

Carboxypeptidase, like TIM, has a solid core of β-structure, clearly visible in Stereoslide 1.

THIS SLIDE IS REQUIRED FOR SAQS 4 AND 5 (OBJECTIVE 2), WHICH YOU SHOULD NOW TRY.

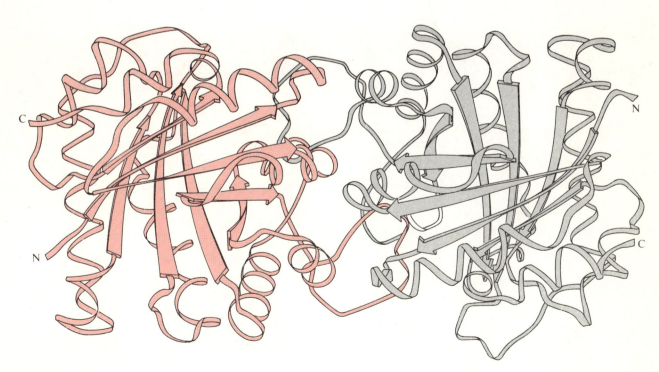

Figure 11 Diagram of tertiary structure of TIM. This shows the dimer, composed of two identical subunits, each a single polypeptide chain. One subunit is coloured red, to emphasize the extensive overlap at the subunit binding site. Features of secondary structure are the eight β-pleated sheets (broad arrows) and the outer α-helices. N indicates NH_2-terminus, and C indicates COOH-terminus of the polypeptide chain.

Note that the model is built from the same kit as in Slide 4. Turn to the description of this if you need reminding what the component parts are. The main information needed, however, is in the H bonding (red).

2.2.2 Folding for maximum efficiency of function

Now that we have described some of the principles behind stability of tertiary structure, we can turn to protein function. In practice, this means seeing how a particular folding pattern produces the various binding sites needed for catalysis (enzyme proteins), regulation (allosteric proteins), transport (carrier proteins) or whatever binding property is required for the function of that particular protein.

binding sites

So far, enzymes are still the most intensively studied group of proteins, and here the principle binding site is the *active site*. This region is designed to bind the substrate strongly and specifically, so that its susceptible bonds come to lie directly in the line of fire of catalytic amino acid residues on the enzyme. In the active sites you will come across both here and on television, you should therefore look out for two features—specificity, and orientation of catalytic groups.

active site

Specificity obviously begins with size—a large substrate cannot be accommodated in a small active site pocket. In TIM, with its single rather small substrate, the active site is a small pit, while in enzymes like ribonuclease or lysozyme, which attack the middle of long chains (see Table 1), it is an extended crevice. Lysozyme, for example, has specific binding regions for six sugar residues.

NOW STUDY STEREOSLIDE 6, WHICH COMES FROM THE TV PROGRAMME ON LYSOZYME.

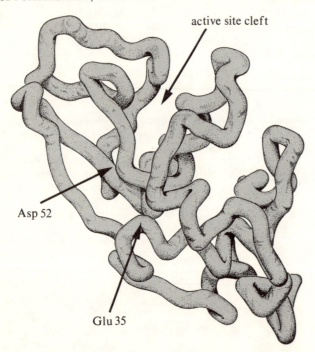

active site cleft

Asp 52

Glu 35

Figure 12 'Sausage model' of lysozyme, showing active site cleft.

Use Figure 12 and the sausage model of Stereoslide 7 to locate where the active site is in Stereoslide 6. Also, do not be confused by the colour coding, which is quite different from that in your building kit. Slide 6 distinguishes between amino acid type (negative, hydrophobic, etc.) rather than atom type (N, C, O, etc.). Tryptophan residues can be identified by their purple (for hydrophobic) aromatic rings. Trp 62 and Trp 63*, both involved in substrate binding, can be seen towards the top left of the cleft, where the first three rings of the hexasaccharide substrate (A—B—C—D—E—F) bind. The lower part of the substrate, including the susceptible bond between rings D and E (see Table 1), is bound in the middle-to-lower part of the cleft. Here you can see the two catalytic groups, Asp 52 higher up on the left and Glu 35 on the right, identifiable by their red colour (coding for negatively charged groups).

Note also the disulphide bond (in yellow) at the top of the molecule behind the active site cleft, holding together separate lengths of polypeptide chain.

* Accepted abbreviations for amino acids, such as Arg, Asp, Glu, Trp and Tyr, can be found in Table 1 of the *Source Book*.

18

One group of enzymes with very large active sites are the coenzyme binding proteins, one of which—lactic dehydrogenase—is shown with its coenzyme in Stereoslide 12 (this will be described in Unit 7). These enzymes* may perform totally different catalytic activities from one another and have different primary structures, but because they use the same coenzymes you would expect their tertiary structures to contain 'domains' of similar shape. The remarkable thing is that these domains may arise from the folding up of quite different polypeptide chains. The evolution of such domains is a very controversial matter.

dehydrogenases

Now let us look more closely at the active site of one enzyme, carboxypeptidase, concentrating on both specificity and orientation of catalytic groups. The enzyme can hydrolyse amino acid residues only from the carboxyl-terminal end of the chain where there is a free COO^- group (see Table 1). Peptide bonds in the middle of the chain or at the amino-terminal end are not attacked. A high resolution picture of the enzyme and its substrate** is now available from X-ray diffraction work, and this gives at least a partial explanation of these specificity requirements. Firstly, the inability to tackle peptide bonds in the middle of the chain is obvious from the size of the active site—a small pit rather than a long crevice. The need for a free COO^- group is less obvious until we follow the sequence of events which take place on substrate binding. The first event is thought to be an attraction between COO^- on the substrate and a positively charged arginine, residue 145 on the enzyme. As this arginine is drawn out of its usual position, a whole series of small conformational changes is set in motion which culminates in a major 1.2 nm shift of Tyr 248, a key residue in the catalytic event. This swings in through 120° and comes to lie close to the vulnerable peptide bond of the substrate.

carboxypeptidase

NOW LOOK AT STEREOSLIDES 1 AND 2, TO COMPARE THE RELATIVE POSITIONS OF TYROSINE BEFORE AND AFTER THE CONFORMATIONAL CHANGE. (Tyr 248 is the purple aromatic residue prominent at the top of Slide 1.)

Evidence for this large conformational change is given in SAQ 14, which you will come to much later. The important point here is that we now have a particularly clear illustration of the induced fit† theory of enzyme catalysis. With carboxypeptidase it is a substrate-induced conformational change that brings the catalytic groups on the enzyme into correct alignment. Without the correct substrate (i.e. one with a free COO^- group) there can be no hydrolysis. Hence specificity can be at least partially explained by a detailed knowledge of groups in the active site pocket.

Non-catalytic binding sites

In none of the proteins just described does the enzyme active site occupy more than a small portion of the total protein surface. What then is the rest of the molecule doing and why are proteins so large? There are two quite different answers, the first of which goes back to the question of stability.

QUESTION Do the amino acid residues in an enzyme active site all come from the same stretch of polypeptide chain?

ANSWER No. Residues close together in the tertiary structure (for instance, those lining the active site pocket) may come from widely separated regions of the primary structure, as you can see in Figure 3 of Unit 1. Obviously, therefore, even a small active site pocket needs a minimum polypeptide chain length if it is to be formed in this way.

Another reason why proteins are so large is that more than one specific binding site may have to be accommodated on the surface. A number of such sites are listed in Table 2. Subunit interaction sites may be merely for physical attachment of the different subunits in the protein, as in TIM (Figure 11). Alternatively, they

---

* These enzymes include the kinases, which need ATP, and the dehydrogenases, which need FAD or NAD. As you can see from formulae in the *Source Book*, all these coenzymes carry the same bulky adenine–ribose–phosphate moiety.

** A substrate analogue rather than true substrate was used. This was the dipeptide glycyltyrosine, a competitive inhibitor which binds in the same position as true substrate. It was diffused in through the crystal, and the electron density map (see Section 2.8) of enzyme plus inhibitor was compared with that of enzyme alone.

may form channels for the relay of conformational changes as in allosteric proteins. An allosteric protein is one whose biological activity at one site A can be altered by events at a remote site B. Sites A and B may be closely similar or identical, as in haemoglobin, or they may be very different, as in allosteric enzymes[A] such as aspartate transcarbamylase (ATCase). When the second site, B, is very different from A, it is usually described as a regulatory site. You will be hearing a great deal more of this in Unit 6; for the moment, we shall concentrate only on haemoglobin and its subunit interaction sites (Figure 13).

allosteric proteins

TABLE 2 Examples of non-catalytic sites on proteins

Site	Protein*
binding site for passenger molecule, e.g. O_2	carrier protein, e.g. Hb, membrane permeases
regulatory site	allosteric enzymes (e.g. ATCase), carrier proteins, etc.
subunit interaction site	proteins with multiple subunits, e.g. ATCase, Hb, PDC, TIM
activation site	zymogens, e.g. proinsulin, chymotrypsinogen
membrane attachment site	may be present in proteins requiring transport to extracellular destination

non-catalytic sites

* Hb = haemoglobin; ATCase = aspartate transcarbamylase; PDC = pyruvate dehydrogenase complex (see Unit 1); TIM = triosephosphate isomerase.

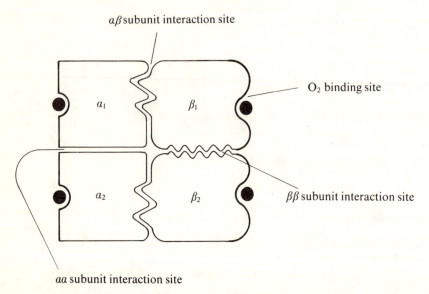

$\alpha\beta$ subunit interaction site

O_2 binding site

$\beta\beta$ subunit interaction site

$\alpha\alpha$ subunit interaction site

Figure 13 Diagram to show variety of specific binding sites in haemoglobin. Note (i) the four almost identical O_2 binding sites and (ii) the three types of subunit interaction site, $\alpha\alpha$, $\alpha\beta$ and $\beta\beta$. (Not shown are the diagonal subunit interactions, $\alpha_1\alpha_2/\beta_2\beta_1$, important in deoxyhaemoglobin.)

Haemoglobin[A] has four subunits (quaternary structure $\alpha_2\beta_2$), each of which has one oxygen binding site. In accordance with its role as an O_2-carrying protein in the blood, the total oxygen binding capacity of the molecule changes with environment—high in the lungs for maximum loading up and low in the tissues for maximum delivery. The mechanism behind this fine adjustment involves a cooperative conformational change relayed from the subunit which binds the first O_2, to the other three subunits. As you might expect, anything which modifies the subunit interaction sites will therefore affect the oxygen-carrying ability of the whole molecule. The clinical effects of this molecular change are considered in ITQ 7.

haemoglobin

ITQ 7 A patient was admitted to hospital suffering from clinical symptoms related to impaired haemoglobin functioning (e.g. mild haemolytic anaemia). Amino acid sequence work on his haemoglobin revealed a genetic defect in which Tyr 35 of the two β-chains had been replaced by a phenylalanine. Compare the formulae of these two amino acids in the *Source Book* and suggest which type of weak bonding may be deficient at the subunit interaction sites of the abnormal haemoglobin.

Finally we come to the zymogens and their activation sites. Many proteins, especially those of potentially hazardous biological activity, are first produced in an inactive form known as a *zymogen*. Activation takes place by proteolytic cleavage at specific activation sites. Common examples are degradative enzymes like pepsin and trypsin, and hormone precursors like proinsulin (Figure 14). The discovery of proinsulin in 1967 removed an embarrassing anomaly from the literature on protein folding. Up till then, denaturation/renaturation studies on insulin failed to regenerate more than a fraction of the expected activity. With hindsight this is hardly surprising, since insulin has lost a substantial length of its original primary structure. In the absence of the complete polypeptide chain, the dogma concerning hydrophobic bonding and free energy cannot apply. *In vivo* the tertiary structure of active insulin is presumably formed in two stages, folding of the polypeptide chain followed by excision of the intervening peptide. To avoid complete proteolytic degradation the protease responsible for activation must be specific for certain features—activation sites in fact—on the proinsulin molecules.

zymogen

activation site

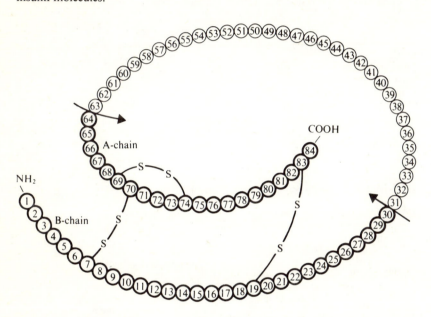

Figure 14 Primary structure of proinsulin. Arrows indicate sites of protease attack. Insulin A- and B-chains (heavy outline) are formed from residues 64–84 and 1–30 of proinsulin. They are held together by disulphide (S—S) bonds, while the excised peptide (residues 31–63) is lost.

Activation sites may therefore be described as regions particularly susceptible to hydrolysis by specific proteolytic enzymes. In this case the partially hydrolysed macromolecule—insulin—is protected from further degradation by its tertiary structure, which appears to offer no further foothold for the activating enzyme.

NOW YOU COULD TRY SAQ 6 (WHICH REQUIRES STEREOSLIDES 6 AND 7) AND SAQ 7.

2.2.3 Change in protein conformation

The protein structures we have just described may leave you with the impression that these molecules are static. In reality, as we described in Unit 1 (Section 1.3.2), they are highly flexible because of the high proportion of weak bonding in their structure. This property is exploited in many biological control mechanisms which operate through *ligand-induced conformational changes*.

A ligand is any small molecule with a specific binding site on the protein, and its effect is to stabilize a conformation with slightly different biological properties —either more active or less active—than the one that predominated before. Carboxypeptidase, described earlier, is a good example of this. Here the ligand is a molecule of substrate, and the conformational change is one which brings

ligand-induced conformational change

catalytic amino acid residues into position in the active site. Many other enzymes show subtle substrate-induced changes in the orientation of active site groups, and this is what lies behind the induced fit theory[A] of enzyme catalysis.

These conformational changes are strictly local, affecting only groups in the region of the ligand binding site. Many proteins have more than one binding site. In the more sophisticated of such proteins, events at one binding site may be relayed through a series of cooperative conformational changes to a more distant binding site in another part of the molecule, often on another subunit. Allosteric proteins are a good example of this, and the effect of such conformational changes on their activity will be discussed in Unit 6. Here we are more concerned with structure, and would like to know how a conformational change is propagated across a molecule. The making and breaking of weak bonds is a key feature of all conformational changes, as we described in Unit 1, but the precise amino acid residues concerned have been only tentatively identified even in haemoglobin, where X-ray crystallography has permitted two conformations —oxy- and deoxyhaemoglobin—to be compared directly.

In a nucleic acid like DNA, which has a large proportion of regular H-bonded structure, it is easier to see how conformational changes may be relayed from one residue to another, and we shall touch on this when discussing nucleic acid structure in Section 2.3.

2.3   Secondary structure in nucleic acids

Study comment This Section requires a basic understanding of the principal dogma of protein synthesis,[A] and in particular of the role of tRNA.[A] If you are unclear about these you should look up the references given in Table A1. Unit 11 of this Course covers nucleic acid function, so here we only give you the necessary background in nucleic acid structure. The concepts of T_m and hybridization (Section 2.3.2) will be assumed knowledge for Unit 11. Otherwise, base pairing (Section 2.3.1) is the most important part of this Section.

Throughout this Section you will need the *Source Book* for formulae of nucleotides and their constituent bases. For the beginning of Section 2.3.2 you will need Stereoslides 8 and 9.

Table 3 should be a useful summary of the secondary structures described here and in the previous Section on proteins.

Our knowledge of nucleic acid structure is not substantiated by as many detailed X-ray crystallographic studies as for the globular proteins, and only one nucleic acid—tRNA—has been studied in comparable depth. This is because tRNA is in many ways more like a protein than a nucleic acid. It is small, with a single 3-D conformation, and it can be crystallized. For this nucleic acid you should therefore be able to relate structure to function in the same way as for proteins (Objective 3b).

Most of the other nucleic acids are very much larger than globular proteins and, as we shall see in the next Section, may assume more than one conformation. Despite this you should be able to pick out certain regular H-bonded features comparable to the α- and β-structures of proteins. These are the double helix of DNA and the hairpin loop in RNA (see Table 3). More elaborate folding patterns —still under the heading of secondary structure—which incorporate the hairpin loop, are seen in the cloverleaf of tRNA. Base pairing (Figure 15) is a key feature of this Section.

2.3.1   Base pairing and base stacking

One of the most important driving forces behind specificity of interaction in nucleic acids, is *base pairing*. This is the principle by which two opposing strands of nucleotides pair up so that the maximum number of strong H bonds can be formed between opposing bases. The 'allowed' base pairs, A-T, A-U and G-C,* produce more H bonds than the alternative disallowed pairs such as A-C or G-T. Furthermore, these bonds would be weaker because the three combining atoms would be less perfectly aligned.

base pairing

* We use the accepted abbreviations A = adenine, T = thymine, U = uracil, G = guanine, C = cytosine. Full structural formulae are given in the *Source Book* under nucleotides.

22

TABLE 3   Secondary structure in nucleic acids and proteins

Structure	Where found	Stabilizing weak bonds
double helix	most DNA;	hydrophobic and H bonds
	RNA of certain viruses;	
	DNA–RNA hybrid helices; single stranded RNA, along the stems of longer hairpin loops	
single helix	single stranded RNA	hydrophobic bonds
hairpin loop	tRNA and probably other types of single stranded RNA	hydrophobic and H bonds
α-helix	globular proteins (in short sections);	H bonds, parallel to chain
	fibrous proteins (e.g. keratin in hair)	
β-pleated sheet	globular proteins;	H bonds, perpendicular to chain
	fibrous proteins (e.g. silk)	

Here we shall consider two secondary structures that result from this base pairing, the *duplex or double helix* found in double stranded DNA and in DNA–RNA hybrids, and the *hairpin loop*. This is found in single stranded RNA, where the chain doubles back on itself. The base pairing principle is exactly the same, of course, whether complementary bases come from the same nucleotide strand (as in the hairpin loop) or from different ones (as in the double helix).

Like other macromolecules, nucleic acids owe their strength initially to a covalently bonded backbone. But both their specificity of interaction and their stability of secondary structure depend on the very large numbers of weak bonds needed to maintain higher orders of structure in what would otherwise be a flexible chain of no fixed conformation. The DNA double helix, for example, is held together partly by hydrogen bonding and partly by van der Waals bonding. The hydrogen bonds, although a comparatively minor factor in terms of total binding energy, are what make for specificity. This is because the geometry of the double helix allows bases to pair with each other only in the ways shown in Figure 15, i.e. A with T, and G with C. Stability can also be influenced by the nature of H-bonded partners, as you will see when we come to describe thermal denaturation.

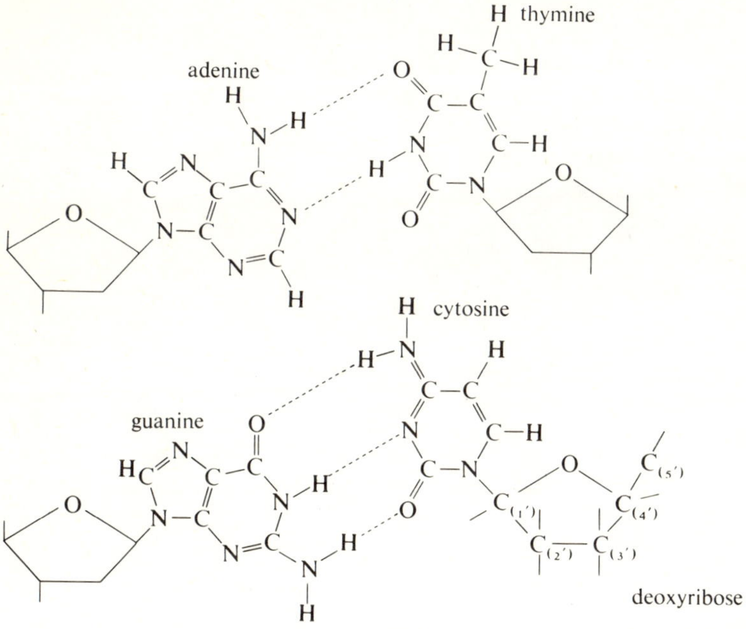

Figure 15   Base pairing in DNA; ···· represents H bonding.

The van der Waals bonds contribute to what is descriptively known as *base stacking*. This may be defined as the hydrophobic interaction between bases lying parallel to each other in any regular helical arrangement. The flat aromatic† rings responsible for base stacking are a prominent feature of the double helix, as you can see from Figure 16. Here their contribution to the total binding energy is reinforced by H bonding, but in single helices the base stacking forces

base stacking

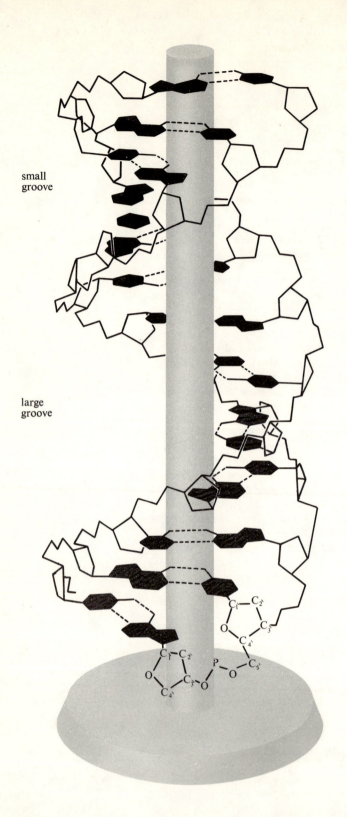

small
groove

large
groove

Figure 16 B form of DNA double
helix.

stand alone. These *single helices* are a well known and fairly stable feature of   **single helices**
secondary structure in single stranded RNA, and are also found *in vitro* in
single stranded DNA.

2.3.2   The DNA double helix

A model of the double helix, as described by Watson and Crick (1953), is shown
in Figure 16 and Stereoslide 8. This is the B form of DNA, which was the form
studied by Watson and Crick. The A form can be obtained by preparing the
fibre under slightly modified conditions. It differs in several important respects
from B-DNA, and it probably has biological significance during transcription
since it is the form assumed in DNA–RNA hybrids.

24

QUESTION Compare the formulae of deoxyribonucleotides and ribonucleotides in the *Source Book*, and suggest one reason for this difference in duplex form.

ANSWER DNA is 'deoxy' because it has lost the bulky O atom from position C-2′ of the ribose ring.*

Loss of this C-2′ oxygen permits the sugar rings to approach more closely than in RNA (or DNA–RNA hybrids). In this way a slight change in primary structure (ribose to deoxyribose) has a major effect on overall shape.

NOW COMPARE THE MODEL OF B-DNA (STEREOSLIDE 8) WITH THAT OF RNA (STEREOSLIDE 9).

The DNA model is built from the same kit as the protein α-helix of Slide 5, so that the relative size of protein and nucleic acid secondary structures is very obvious.

QUESTION How many turns of double helix are there in the length of RNA shown in Slide 9?

ANSWER Exactly one. (Note that Slide 5 has approximately three turns of α-helix in a very much smaller space.)

QUESTION Compare Stereoslides 8 and 9, and suggest what are the two major differences between the A and B forms of double helix. (Note that DNA is built from a kit in which black spheres represent carbon, red spheres oxygen, *small* white spheres hydrogen, and *large* white spheres phosphorus.)

ANSWER (i) In form B the base pairs lie flat, perpendicular to the main chain axis, while in form A they lie tilted at approximately 20° to it.

(ii) Twisting of the main chain produces two grooves along the outer edge of the molecule—a large and a small one in B, and two equal-sized ones in A. (See also Figure 16.)

These grooves are thought to accommodate histones and other proteins associated with nucleic acid in the living cell in higher organisms. This association may be one factor controlling the shape of the DNA duplex, and hence the availability of specific regions for transcription. So once again there may be biological significance in these apparently minor structural details. In neither Slide 8 nor Slide 9 can the identity of bases be made out, but the structure of ribose (or deoxyribose) sugar backbone is very clear. Using the sugar formula given in the right-hand corner of Figure 15, you should be able to make out from the stereoslides the sugar ring O (in red), and the 3′ and 5′ carbons which provide the link points in the polynucleotide chain. Note also the C-2′ position. Here RNA has a bulky OH (not shown) that forces the helix into the A form.

QUESTION The top of Slide 9 shows the beginning of two nucleotide chains. Take the *lower* of these. Reading from top to bottom, does it run 5′ → 3′ or 3′ → 5′?

ANSWER It runs 3′ → 5′, reading from the top downwards. To locate yourself on the sugar ring, note that C-1′ is where the base substitutes in, that the numbering goes clockwise from there and that C-5′ is not in the sugar ring itself, but occurs as a substituent on C-4′. It is then apparent that C-3′ branches, through phosphate, to C-5′ of the next residue, i.e. the chain, running downwards, reads 3′ → 5′.

The two strands of double helix are held together by weak bonding. Therefore the critical processes of winding and unwinding—both relevant to replication—can readily be controlled without any major input of energy. During replication, a partially unwound duplex with its two strands of exposed nucleotides acts as template for the formation of two daughter strands. The exposed unpartnered

* To avoid confusion, ring positions on the bases are numbered 1–6 (or 1–9 for the purines), while ring positions on the sugar are numbered 1′–5′.

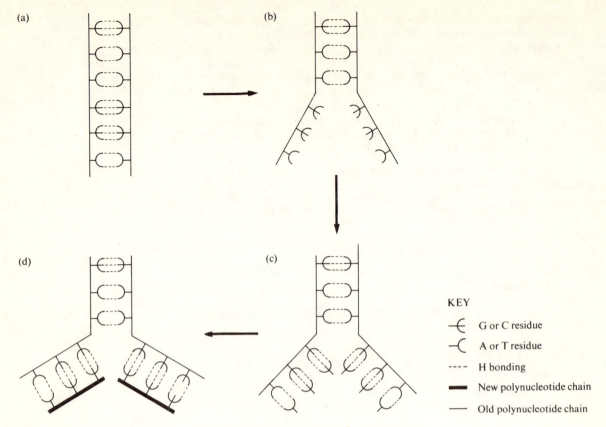

(a)

(b)

(d)

(c)

KEY

G or C residue

A or T residue

---- H bonding

New polynucleotide chain

Old polynucleotide chain

Figure 17 Role of complementary base pairing in DNA replication. (a) Strands of parent double helix, held together by H bonds (····) between complementary base pairs. (b) Partial strand separation. (c) 'Free-floating' bases from surrounding medium attracted to unpartnered bases on exposed DNA strand, by H bonds between complementary base pairs. (d) Newly-aligned bases joined together by covalent phosphodiester bonds in backbone chain.

nucleotides will rapidly form H bonds with complementary bases from free tri-nucleotides in solution (see Figure 17), and these will be 'zipped' together into a polynucleotide chain by the action of polymerase or ligase enzymes. The important point is that of the many random H-bonded duplexes which might result, only the most energetically stable one is found. This is the one in which, for reasons of geometry explained earlier, the bases on the new strand are exactly complementary to those on the original template. Hence much of the faithfulness of replication can be explained in terms of weak bonding between complementary base pairs.

replication

Thermal denaturation as an example of cooperativity

In vitro, newly formed pairs of polynucleotide chains are thought to twist spontaneously into a helix. This process can be followed in reverse during the unwinding which precedes strand separation when DNA is *thermally denatured*. This requires heating to at least 50 °C and is therefore not a physiological process, but we shall include it here because it forms the basis of hybridization—a technique which has been extremely useful in establishing relationships between different DNAs.

thermal denaturation

The unwinding (or winding) of DNA is another example of a *cooperative conformational change*, i.e. one in which a whole series of far-reaching changes are generated from one initial event. With proteins this initial event is often the binding of a regulatory ligand, as we described in Section 2.2.3. With nucleic acids the initial event *in vivo* may be binding of polymerase or even a specific 'unwindase' protein, while in thermal denaturation *in vitro* it is simply local H-bond breaking induced by heating. This bond breaking does not remain local, because in a macromolecule it is difficult for one residue to move without disturbing its neighbours. So any local conformational change is propagated all along the line. Hydrogen-bonded elements of regular secondary structure are particularly prone to this.

cooperativity

26

Figure 18 shows what is known as a melting curve for DNA. As you can see, the unwinding process is followed indirectly by a property which is dependent on the amount of secondary structure in the molecule. Both optical rotatory dispersion (Section 2.7) and absorption of ultraviolet light (Section 2.6) are particularly suitable for this. From the melting curve can be calculated T_m, a physical constant much used in characterizing different DNAs. It may be defined as the temperature needed for 50 per cent denaturation, and is calculated from the midpoint of the steep part of the curve. The T_m value can give information on two aspects of nucleic acid structure: degree of double-strandedness and, as you can see in ITQ 8, base composition.

melting curve

T_m value

> ITQ 8   DNAs with a high proportion of G+C residues have high T_m values compared to those with a high proportion of A+T. Suggest why this should be. *Hint*   Look at Figure 15.

A melting curve can also predict whether a nucleic acid solution contains double or single stranded molecules.

> ITQ 9   Compare the shapes of curves A and D in Figure 18 and suggest two features which might be used to distinguish between single and double stranded DNA.

Steepness of slope and the presence of plateau regions at either end of the curve are both features common to all the double stranded DNA curves, and missing in the single stranded DNA curve. This is because they are characteristic of a highly cooperative process. The lower region shows that double stranded DNA resists unwinding over a considerable temperature rise, until a critical point is reached. Unwinding is then transmitted rapidly through the molecule, for very little further rise in temperature. Cooperativity will be lacking in a single stranded molecule which has no H-bonded secondary structure, and therefore the melting curve D is much flatter, indicating less abrupt unwinding of a single rather than a double helix.

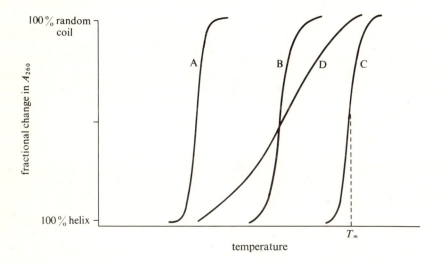

Figure 18   Melting curves observed on heat denaturation of DNA. T_m (shown only for curve C) is the temperature for 50 per cent denaturation. A, B and C are double stranded DNAs of increasing G+C content. D is a single stranded DNA.

Hybridization

A technique that exploits this difference in double and single stranded behaviour is hybridization. This is a very popular method for estimating relationships between nucleic acids without the enormous labour of sequencing them (remember that even bacterial DNA has a molecular weight some 10^5 times greater than that of the average protein). In complete denaturation, the two DNA strands are first unwound and then separated. In renaturation, careful cooling reverses this process, the first step being for the strands to come together again, attracted by complementary base pairing. Rewinding then follows spontaneously. If, during cooling, another (single-stranded) nucleic acid of unknown sequence is introduced into the solution, the DNA may pair with this rather than with its original partner. Whether or not the new *hybrid* double helix is stable depends solely on the degree of complementarity between the base sequences, and this is the principle behind all hybridization experiments. You may judge their usefulness in solving the problems of biology from the next question.

hybridization

QUESTION Single stranded DNA is allowed to renature in the presence of mRNA which is thought to have been transcribed from it. Would you expect the renatured solution to contain mainly double stranded or mainly single stranded molecules,

(a) if the RNA is indeed a transcription product of the DNA,

(b) if it has been transcribed from some other DNA and is therefore not related to it?

ANSWER (a) If the RNA has been transcribed from the DNA it will be exactly complementary to it, and the renatured solution will contain mainly double stranded molecules—both original DNA duplex and hybrid DNA–RNA duplices.

(b) If the RNA is not related to the DNA it will be unable to base-pair with it. Furthermore it may physically impede strand recognition between original DNA partners, and the solution will contain mainly single stranded molecules.

NOW YOU COULD TRY SAQS 8 AND 9.

2.3.3 Secondary structure of single stranded RNA

RNA rarely takes up the double helical structure we have just described, except in certain viruses which use RNA instead of DNA as a store of hereditary information (see Unit 11). Most cellular RNA is concerned with the business of translating this hereditary information into protein and is not double stranded.

Messenger RNA (mRNA) is formed on the DNA template by transcription and then bound to ribosomes for translation. These particles also contain RNA—the 5S,* 16S and 23S *ribosomal* RNAs, all of which are closely associated *in vivo* with specific ribosomal proteins. A much smaller molecule with a molecular weight of around 25 000 is *transfer* RNA (tRNA), which has an adaptor role in 'transferring' amino acids from the cytoplasm to the correct site on the messenger. This adaptor role of tRNA requires first that a covalent bond be formed between tRNA and amino acid, forming a key intermediate, the *aminoacyl tRNA*. This reaction is catalysed by a specific *synthetase* enzyme which has two binding sites, one for tRNA and one for the amino acid corresponding to it, i.e. the amino acid coded for by the trinucleotide sequence on the mRNA (codon), which base-pairs with the complementary trinucleotide sequence (anticodon) on the tRNA (see Figure 19).

mRNA

rRNA

tRNA

synthetase

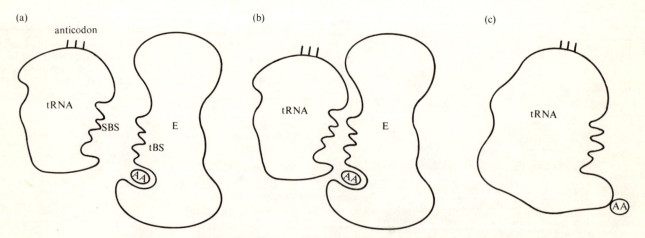

Figure 19 Adaptor role of tRNA. (Any induced-fit conformational changes have been ignored, and relative sizes of tRNA and synthetase are approximate only.) (a) Synthetase and uncharged tRNA before interaction. AA represents amino acid bound non-covalently in the specific amino acid binding site, SBS represents synthetase binding site on tRNA, and tBS represents tRNA binding site on the synthetase. (b) E—S complex in which E (synthetase) catalyses formation of a covalent bond between its two substrates (tRNA and amino acid). (c) Release of tRNA charged with specific amino acid.

* S is the Svedberg unit described in Unit 1. These are the values found from bacterial rRNA. Higher organism rRNA differs slightly.

Many of the smaller RNA molecules have now been analysed and their nucleotide sequences are known with some certainty. What is not known, except for inspired guesswork, is how these single polynucleotide chains fold up into higher order structures like the DNA duplex. Figures 20 and 21 show some of the folding patterns which have been suggested. The principle behind the inspired guesswork is to fold so as to maximize the number of H bonds which can be formed by complementary base pairing. As you see, this produces hairpin loops at the ends of what may be quite long stretches of base pairing. These base-paired sections resemble the double helix of DNA, and it is very probable that they too twist into a helix. As you may remember, the DNA duplex needs only ten bases to form one turn of helix; the advantage of such an arrangement is that helical segments are further stabilized by hydrophobic bonding from base stacking.

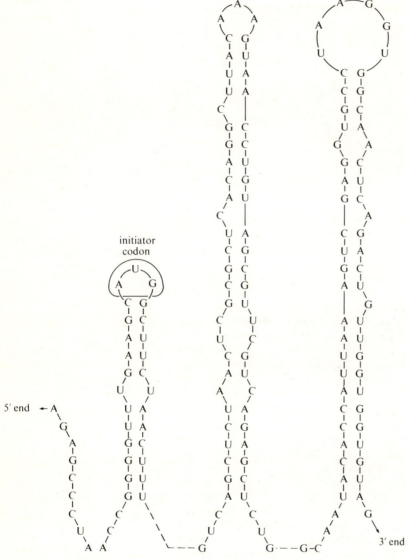

Figure 20 Viral RNA. Postulated secondary structure for part of the single stranded RNA from a virus that uses RNA both as a hereditary information store and as mRNA.

ITQ 10 The nucleotide sequence of an imaginary length of RNA is:

A-U-U-U-A-A-A-U-U-A-A-A-A-A-U

1 2 3 4 5 6 7 8 9 10 11 12 13 14 15

(a) Arrange this sequence into a hairpin loop with maximum opportunities for base pairing.

(b) How many intrachain H bonds are then available for stabilizing this structure?

(c) Is the base-paired section long enough for base stacking forces to contribute to stability?

The tRNA cloverleaf

With tRNA, the suggested secondary structure is more than just speculation and there is now substantial experimental evidence for a cloverleaf arrangement. This

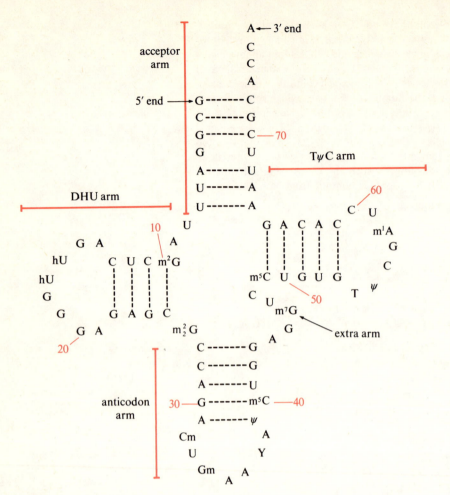

Figure 21 Cloverleaf diagram of the secondary structure of yeast tRNA$_{phe}$.

is based mainly on the accumulated data of a large number of nucleotide sequences from different tRNAs. All these, whatever their primary structure, can be folded into the same basic cloverleaf pattern, and in each the key anticodon region is prominently exposed. (More recently, further evidence has come from X-ray work on tRNA$_{phe}$,* which we shall be describing under tertiary structure.) One great attraction of the cloverleaf model is the way it relates structure to function, by so prominently displaying the three key nucleotides on the anticodon arm. How far can structure be related to function for the other arms of the cloverleaf? Figure 21 shows five such arms—the acceptor, anticodon, TΨC,** DHU and extra arms. The acceptor arm contains the two free ends of the molecule, but the other four arms are continuous, forming four hairpin loops. The stems of all five arms are stabilized along their lengths by H bonding between complementary base pairs, while the four loops at the ends (where the 'unusual' bases tend to congregate) are free. The DHU and TΨC arms are named after the two bases, dihydrouracil (DHU) and pseudouracil (Ψ), which are always to be found somewhere in this region of the molecule. The TΨC arm is thought to be concerned in ribosome binding. The extra arm is so named because it was 'extra' to the cloverleaf originally depicted from the sequence of tRNA$_{ala}$, where it is only two nucleotides long and therefore escaped notice. In some tRNAs it runs to 14 nucleotides but, like the DHU arm, it cannot yet be assigned to any specific function. In contrast, the acceptor arm has a very obvious function, for here is the free end to which the amino acid attaches. This is the reaction catalysed by aminoacyl tRNA synthetase. The specific interaction sites

---

* The subscript phe indicates that this is the tRNA for the amino acid phenylalanine. Similarly, the subscript ala refers to alanine.

** There is at least one tRNA for each amino acid, and one way this individuality is made possible is that several bases other than the four common ones occur throughout the sequence. The formula of one such, pseudouridylic acid, is given in the *Source Book*. Note particularly pseudouracil (symbol Ψ, psi), dihydrouracil (DHU), dimethylguanosine, and 5-methylcytosine, all of which are found quite frequently (see for example Figure 21).

by which these two macromolecules—tRNA and synthetase—recognize each other must be carefully tailored (see Figure 19). Otherwise, there will be mis-matching between tRNA and amino acid, leading to the insertion of the wrong amino acid into the growing polypeptide chain. The synthetase binding site on the tRNA will include nucleotide residues from the acceptor arm, but contributions from nucleotides on other arms of the cloverleaf cannot be discounted. The fully formed tertiary structure, just as in proteins, may juxtapose residues from quite distant parts of the extended chain.

In tRNA, the proposed secondary structure folding can therefore be rationalized in terms of function. With other types of RNA, like the viral RNA of Figure 20, the folding pattern is still speculative but some sections of it can tentatively be related to function. Initiator codons,* for example, tend to be found in exposed positions at the ends of hairpin loops.

NOW YOU COULD TRY SAQ 10.

2.4 Higher order structure in nucleic acids

2.4.1 Do nucleic acids have a unique conformation?

We have just described some of the H-bonded features which may be found in nucleic acids, such as cloverleaf for RNA, and A and B double helices for DNA. You can see already that more than one secondary structure may be formed from the same nucleotide sequence, and when we come to higher orders it is clear that the dogma 'primary dictates tertiary' does not apply to nucleic acids in the same way as for proteins. The only exception is tRNA, a rather small nucleic acid (molecular weight 25 000) whose tertiary structure has recently been analysed by X-ray crystallography. The very fact that tRNA molecules crystallize at all suggests that, like the proteins, they have a unique conformation.

This is not true of most other nucleic acids. Some of these run to molecular weights of several hundred million, and to determine the folding pattern of such an enormous chain is obviously no mean feat. Furthermore, many nucleic acids are *in vivo* closely associated with protein.** This probably has a strong directing influence on the nucleic acid conformation. When studying the conformation of purified DNA, you should therefore bear in mind that in the cell the picture may be somewhat different. This is an area of much active research, and it is early yet to draw firm conclusions. Recent ideas about the higher order structure of DNA are discussed in more detail in Unit 11 and in the TV programme on chromatin.

2.4.2 Tertiary structure of tRNA

Study comment This Section describes the tertiary structure of the first nucleic acid to be crystallized. It ties in some of the principles of conformation—particularly those governing secondary structure of nucleic acids—to the techniques to be described in Part II. For this reason you may not understand all the references to X-ray crystallography until you have read Section 2.8.

Experimental difficulties

The story behind the structure determination of tRNA is a good example of the need to approach a problem of this size from more than one experimental angle. Although the electron density maps that result from X-ray data contain potentially more information than any other method, the interpretation of such maps requires substantial corroborative information from independent techniques like those we are about to describe (Part II of this Unit).

---

* The initiator codon AUG codes for *N*-formylmethionine, an amino acid thought to come at the beginning of all newly synthesized polypeptide chains in bacteria; it therefore represents a 'start reading here' message (see Unit 11).

** The term *nucleoprotein* refers to a non-covalently linked protein–nucleic acid complex. Examples of the protein component are the ribosomal proteins, and the histones which form an integral part of the chromosome in higher organisms.

The first tRNA structure, published in 1973 by an American group (Kim *et al.*, 1973), has since been substantially revised. Figure 22 shows a rather different structure published shortly afterwards by Klug's group in Cambridge (Robertus *et al.*, 1974; Ladner *et al.*, 1975). This is now the generally accepted model; it differs from the first structure mainly in the position of the DHU arm, which lies in a region of high electron density. It was difficult here to distinguish between separate tRNA molecules lying close together in the crystal.

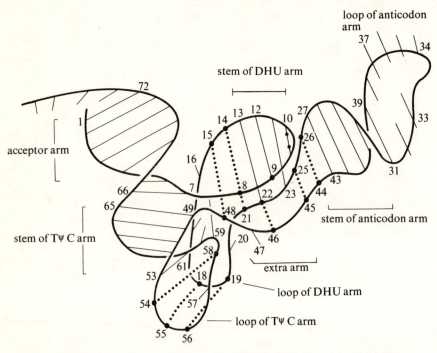

Figure 22 Diagram of tertiary structure of yeast tRNA. The ribose–phosphate backbone is shown as a heavy continuous line. The longer straight lines represent base pairs in the double helical regions, the shorter ones represent single bases in unpaired regions. Dotted lines represent base pairs stabilizing tertiary structure.

The problems of interpreting electron densities can be seen from Figure 23, which shows part of the 3 Å map from which the Cambridge structure was first deciphered. As you can see in Figure 23a, the phosphate groups of the ribose–phosphate backbone are the most electron-dense features of the map. These form the main chain, and once this has been deciphered there comes the problem of assigning electron density peaks to specific nucleotide bases. Several of these are shown in Figure 23b. (Try to imagine the electron density contours before the chemical formulae were superimposed.)

Structural features

The diagram of yeast tRNA$_{phe}$ shown in Figure 22 is based on a more detailed map at 2.5 Å resolution. All the elements of secondary structure can be made out. The TΨC and acceptor arms together form a *double helix* twelve residues long, twisted just as in the DNA duplex. The anticodon loop forms a *single helix* stabilized by base stacking between residues 32 and 33, and between residues 34 and 38.

A surprise feature was the overall shape, an L rather than the X you might have expected for a cloverleaf. The anticodon and acceptor arms are both prominently exposed, but the DHU arm is twisted to lie with its loop pointing downwards, forming base pairs with residues from the TΨC loop. Note, for example, the guanine 19–cytosine 56 pair. This is a typical Watson–Crick pair of the type found in DNA; other parts of the tertiary structure are stabilized by unusual base pairs, e.g. guanine 15–cytosine 48, or by base triples, e.g. cytosine 13–guanine 22–methylguanine 46. These are both shown in Figure 23a.

It is thought that this structure will accommodate all tRNA molecules with a cloverleaf pattern similar to that of yeast tRNA$_{phe}$.

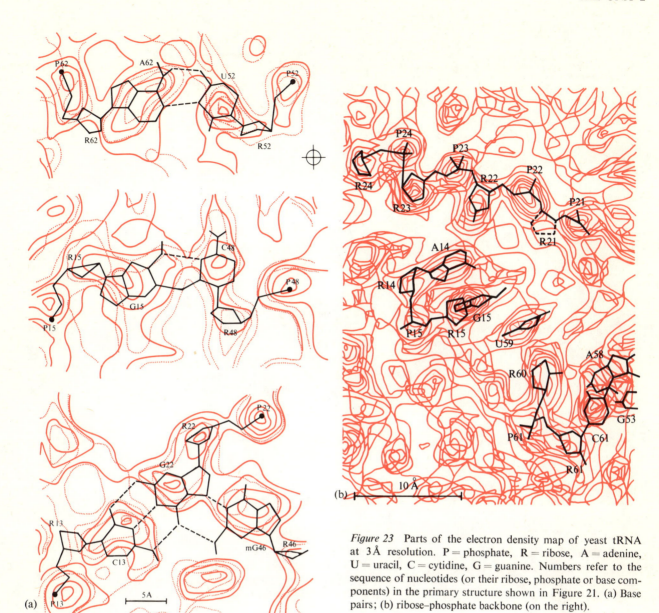

Figure 23 Parts of the electron density map of yeast tRNA at 3Å resolution. P = phosphate, R = ribose, A = adenine, U = uracil, C = cytidine, G = guanine. Numbers refer to the sequence of nucleotides (or their ribose, phosphate or base components) in the primary structure shown in Figure 21. (a) Base pairs; (b) ribose–phosphate backbone (on the right).

TABLE 4   Properties of DNA from various sources

Organisms	Molecular weight of double stranded molecule	Shape
viruses	1×10^6	linear, or occasionally circular (may use sticky ends)
bacteria	$2 \times 10^9 - 4 \times 10^9$	circular, but probably looped and supercoiled around RNA core
higher organisms	8×10^{10}	highly condensed by looping and supercoiling, as part of nucleo-protein complex (see Unit 11)

2.4.3   Higher order structure of DNA

Table 4 lists some of the properties of DNA from different sources. A striking feature is the enormous molecular weight, particularly in the higher organisms where each chromosome consists only of protein plus a single molecule of DNA. *Viral DNA* has the simplest form, a linear helix, but for replication this too becomes circular. This necessary change in conformation may explain the *sticky ends* found in many viral DNAs. These are short lengths (about 12 nucleotides

sticky ends

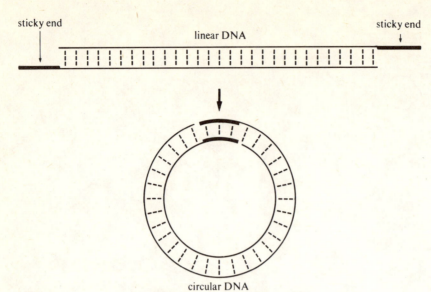

circular DNA

Figure 24 Speculative diagram showing the role of sticky ends in viral DNA. Broken lines represent H bonds between complementary base pairs. The twelve-nucleotide-long sticky ends of the nucleotide chain are shown as a thick line.

long) jutting out at opposite ends of the linear molecule (Figure 24). The base sequences of these apparently distant ends are always complementary to each other.

QUESTION   What weak bonds could be formed spontaneously if the linear molecule were temporarily to assume a circular conformation?

ANSWER   H bonds between complementary base pairs from opposite ends of the molecule. These are shown in Figure 24, where you see that once again base pairing has a directing influence on conformation.

In *bacteria* the DNA is always circular rather than linear. This was very clearly demonstrated in Unit 1 (Figure 27) in the autoradiograph of the replicating *E. coli* chromosome. However, recent work suggests that bacterial DNA spends much of its time tightly supercoiled (see below) and looped around a central core of RNA (see Figure 25).

The enormous lengths of DNA in *higher organisms* must be even more tightly folded, and it is difficult to see how such compact molecules can undergo replication without at least partial unfolding. This could well provide a point of control in the replication of both bacteria and higher organisms, but so far this idea remains highly speculative (see Unit 11).

One way of folding DNA is by *supercoiling*, which may be imagined as the result of forcing a length of linear double helix into a few extra twists before tying its ends into a circle with a covalent bond. The resultant strain causes buckling into supercoils, which you can see in Figures 26 and 27. The precise degree of supercoiling may be controlled *in vivo* by protein binding in the grooves along the outside of the double helix, while *in vitro*, with purified DNA, it can be shown that supercoiling is influenced by *intercalation*. Here the binding molecule is flat and slips in between adjacent base pairs—which are also flat, of course. Ethidium bromide, which not surprisingly is an inhibitor of protein synthesis, is one such example, and the antibiotic actinomycin is another. Figure 28 suggests how binding of intercalator could alter the degree of supercoiling in a DNA helix.

supercoiling

intercalation

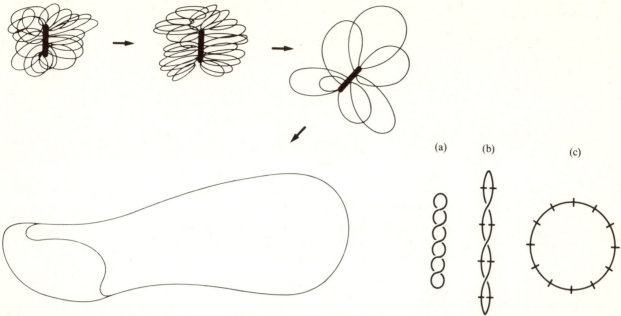

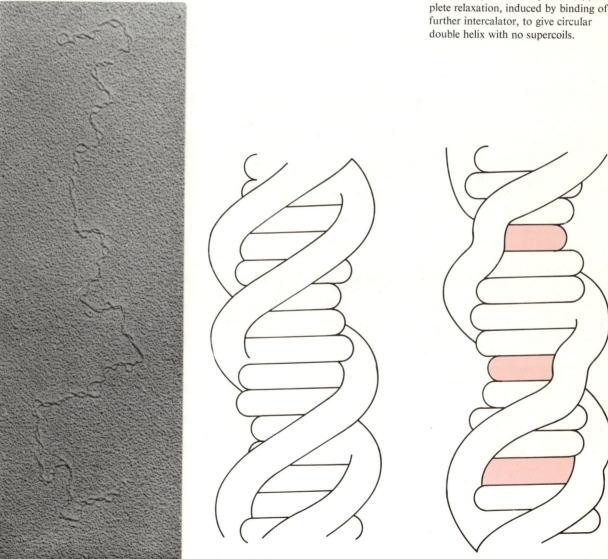

Figure 25 Unfolding of *E. coli* chromosome, as induced by binding of intercalator. Note that the RNA core (thick black line) has dissociated from the fully unfolded chromosome, which now resembles that in Unit 1, Figure 27.

Figure 26 Supercoiling of DNA double helix. (a) Five supercoils; (b) partial relaxation, induced by binding of intercalator, to give three supercoils; (c) complete relaxation, induced by binding of further intercalator, to give circular double helix with no supercoils.

Figure 27 Electron micrograph of bacteriophage DNA, showing supercoils.

Figure 28 Speculative diagram showing change in supercoiling of DNA when intercalator binds. Intercalator molecules (red) insert themselves between base pairs (white), forcing the helix to unwind.

2.5   The range of techniques

As in Unit 1, we do not wish to leave you with a list of principles and no means of assessing the experimental evidence behind them. Therefore, in this part of the Unit we aim to give you an idea of the type of methods used in studying conformation, without going into too much experimental detail. Table 5 summarizes the key points. It should underline for you what is actually being measured, what exactly comes out of these measurements, and how this relates to conformation. It should be a useful guide to refer to while reading through the text, but obviously much of it will be obscure till you have reached the end of the Unit.

Of the many ingenious methods for studying conformation in macromolecules, we shall concentrate on two approaches—optical techniques (i.e. spectroscopy and X-ray diffraction) and chemical modification. Chemical modification has

TABLE 5   Experimental techniques for investigating conformation

Technique	What is measured experimentally	Macromolecule examined*	Information obtained*
UV absorption spectroscopy	1 difference spectrum, i.e. A in presence and absence of ligand, etc.	protein	ligand-induced conformational changes
	2 spectrophotometric titration curve, i.e. A at fixed wavelength for range of pH values	protein with ionizing chromophore, e.g. tyrosine	pK value of individual residues in macromolecule
	3 melting curve, i.e. A during heat denaturation	nucleic acid	T_m value → calculation of % G+C
	4 hypochromism, i.e. observed A compared with theoretical A calculated from constituent nucleotides	nucleic acid of known composition	% helical content; proportion of single- to double-strandedness
fluorescence	fluorescence intensity at λ_{max} in emission spectrum, in presence and absence of ligand	proteins containing Tyr or Trp; nucleic acids; membranes, etc., labelled with fluorescent probes	quenching, indicating ligand-induced conformational change
ORD or CD	1 ORD or CD spectrum compared with that of model compound	nucleic acid	% double-strandedness; % helical content
	2 ORD or CD at fixed wavelength, during heat denaturation	nucleic acid	T_m value
	3 ORD or CD spectrum in presence and absence of ligand, etc.	protein, tRNA	ligand-induced conformational changes
NMR	1 spectrophotometric titration curve, before and after chemical modification of histidines	ribonuclease (details of experiment in Unit 4)	pK value of individual histidine residues
	2 NMR spectrum in presence and absence of ligands	tRNA; globular proteins; membranes	ligand-induced conformational changes
X-ray diffraction by fibres	position of 'spots' in diffraction diagram	fibrous proteins; polysaccharides; DNA	distance between repeat units in the molecule
X-ray crystallography	position and intensity of 'spots' in diffraction diagram, from molecule with and without heavy atom	crystalline globular proteins and tRNAs	electron density map → complete tertiary structure (provided sequence is known)
chemical modification	1 spectrum of modified molecule in presence and absence of ligand, etc.	protein or nucleic acid	conformational changes in regions lacking natural reporter groups
	2 amino acid (or nucleotide) analysis to identify modified residues	protein or nucleic acid	mapping of specific residues as 'inside' or 'outside' in the tertiary structure

* This Table summarizes the examples mentioned in the Unit. It by no means exhausts the type of problem that can be tackled.

been listed separately although, as you will see, it is often used in collaboration with the optical methods. The term 'optical techniques' is a blanket term for all the different ways matter may interact with electromagnetic radiation. We have taken it to include X-ray diffraction where the radiation is scattered by the macromolecule, as well as spectroscopy where it is absorbed.

Many different branches of spectroscopy have been applied to the study of macromolecules, and all these concern the conformation of specific *chromophores*, i.e. those parts of the molecule that absorb incoming radiation at a particular wavelength. Examples of chromophores that absorb in the ultraviolet (around 280 nm) are tyrosine residues in proteins, and the bases (all five of them) in nucleic acids. In Section 2.6.1 we concentrate on just one method, ultraviolet absorption spectroscopy, but many of the ideas described here can be applied to other branches of spectroscopy. We aim to show this in Section 2.7, where fluorescence, optical rotatory dispersion and nuclear magnetic resonance are described briefly. You will notice throughout these two Sections that there are two ways of looking at spectra. One is to focus on specific chromophores and to study local conformational changes in their vicinity. The other is to take an overall view, and study the way in which the *relative* orientation of individual chromophores can influence their interaction with incoming radiation. Where the molecule has lengths of regular secondary structure, this approach can give information about the proportion of secondary structure in a molecule. It has often been used in denaturation studies.

chromophore

The last method to be described is *chemical modification*. This is a means of labelling specific amino acid or nucleotide residues on a molecule, by causing them to undergo some chemical change. It is used for conformational studies in two ways: it may introduce 'probes' or reporter groups into interesting regions of the molecule, which can then be studied by spectroscopy; or it may be used to map the position of residues ('inside' or 'outside') in the tertiary structure.

chemical modification

None of these methods, with the possible exception of nuclear magnetic resonance, have anything like the resolving power of X-ray crystallography. This is the branch of X-ray diffraction concerned with macromolecules that will crystallize, and as you will see in Section 2.8 it is capable of distinguishing between individual atoms in the molecule. Although this technique may seem to be the ultimate in conformational studies, it is by no means true that it is the only one worth pursuing. Not all macromolecules will crystallize for a start and, even when they do, X-ray crystallography can give only a static picture of the molecule. The changes in conformation that take place when the molecule is in action are usually studied by spectroscopy. The essence of a good conformational study, therefore, is to assemble corroborative evidence from as many different techniques as possible. We shall now describe some of these in more detail.

2.6 Ultraviolet absorption spectroscopy

Study comment In this Section we introduce several concepts which are relevant to all types of spectroscopy—difference spectra, spectrophotometric titration and the assignment problem (i.e. the difficulty of assigning peaks in the spectrum to specific residues in the macromolecule). You should remember, therefore, that much of this Section may be relevant to other spectroscopic techniques mentioned throughout the Course.

Some understanding of underlying physical principles is essential, particularly as absorption of radiation occurs in all spectroscopic methods. The relevant knowledge for this covers energy levels, wavelength and aromaticity, all of which are listed in Table A1.

UV absorption is the simplest method of measuring macromolecular concentration in solution. Although this is only indirectly relevant to conformation, no conformational studies can proceed without some idea of concentration, so this Section should not be omitted.

With nucleic acids, UV absorption studies tend to concern conformation of the molecule as a whole, while in proteins the emphasis is on conformation of individual residues. You should be aware of the reasons for this difference, and should be able to give examples showing how conformational changes of both types can be monitored by UV absorption.

Sections 2.6.1 and 2.6.2 are the most important. You may omit the latter part of Section 2.6.3, so long as you are clear what is meant by hypochromism.

2.6.1   Characteristics of the UV absorption spectrum

If a beam of UV light is passed through a macromolecular solution, the chances are that on emerging it will be deficient in radiation at certain wavelengths. The missing radiation has been absorbed by the macromolecule in question, and the precise wavelengths absorbed are a characteristic of that particular molecule. This is because matter interacts only with discrete units or quanta[A] of radiation, and the size of these quanta is determined by the precise arrangement of energy levels[A] within the molecule.

A typical *absorption spectrum* for a protein is shown in Figure 29. This is a plot of the difference in radiation intensity over a given wavelength range, before and after passing through the protein solution. Many modern spectrophotometers compare the intensities of incident and transmitted radiation automatically, by splitting the incident beam into two halves (see Figure 30). One half passes through the macromolecular solution and the other through the 'blank' (this is usually the buffer in which the macromolecule is dissolved).

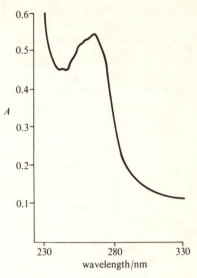

Figure 29   UV absorption spectrum of one component of myosin, the muscle protein.

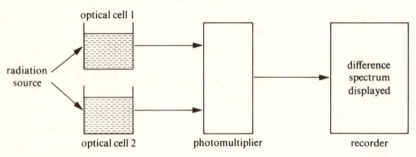

Figure 30   Difference spectroscopy using split beam spectrophotometer. The radiation source emits monochromatic (single wavelength) light, although the precise wavelength emitted at any one time can be varied over a wide range. Identical beams of incoming radiation from this source pass through the two solutions under observation. The difference in transmitted radiation is then displayed directly. A simple *absorption spectrum* results when the two cells contain respectively a protein solution and its 'blank' (see text). A *difference spectrum* results when each cell contains the same protein solution under different sets of conditions.

There are four chromophores which contribute to the UV absorption of proteins: the peptide bond, and the three aromatic amino acid residues, tyrosine, tryptophan and phenylalanine. At low wavelengths the absorption spectrum is dominated by a large peak at 220 nm, as you can see in Figure 29. This is the peptide bond absorption peak, which gives little information about molecular structure because of overlapping absorption peaks from dissolved oxygen and water. At higher wavelengths we come to more useful absorption by specific amino acid sidechains. Aromatic† rings tend to promote absorption in the UV, because of the delocalized electrons in the molecule. The three aromatic amino acids therefore all contribute to the spectrum of the whole protein, although phenylalanine plays only a minor role. The individual spectra for these amino acids are shown in Figure 31, and illustrate the first characteristic of an absorption spectrum, λ_{max} (lambda max). This may be defined as the *wavelength of maximum absorbance*.

absorption maximum, λ_{max}

ITQ 11   What are the λ_{max} for Trp and Tyr under the (unspecified) conditions prevailing in Figure 31?

The reason why phenylalanine plays such a minor part is that, mole for mole, it absorbs much less strongly than either tyrosine or tryptophan. Quantitatively this fact can be expressed in our second characteristic, the molar absorbance ϵ (epsilon), sometimes called the extinction coefficient. This is defined as the *absorbance at a given wavelength of a one molar solution of the molecule in a 1 cm cell* (i.e. in a transparent container 1 cm wide), and is derived from the equation

molar absorbance, ϵ

$$A = \epsilon c l \qquad (2.1)$$

where A is absorbance† (alternatively called the extinction, E, and previously called OD or optical density), c is the concentration in moles per litre, l is the

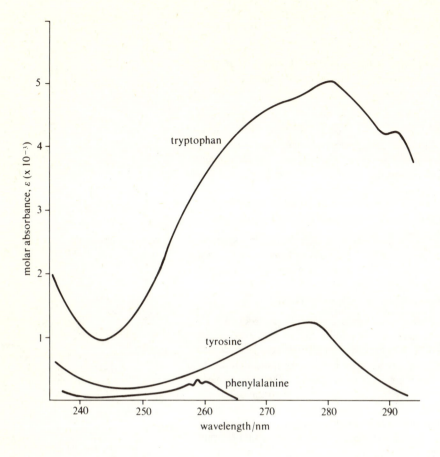

Figure 31 UV absorption spectra of amino acid chromophores.

length of the optical path in centimetres, and ϵ is the molar absorbance. In practice, UV absorption is a very important way of estimating the protein (or nucleic acid) concentration in a solution since, for a given molecule, molar absorbance at the λ_{max} is a constant.

> **ITQ 12** The molar absorbance of the enzyme pepsin is 5.1×10^4 (at 278 nm, the λ_{max} position). If the molecular weight of pepsin is 35 800, calculate:
>
> (a) the concentration, in moles per litre, of a pure pepsin solution where A_{278} (absorbance at 278 nm) is 0.8;
>
> (b) the expected A_{278} value of a 10 mg ml^{-1} solution of pure pepsin. *Hint* A molar solution contains x g l^{-1} or x mg ml^{-1}, where x is the molecular weight.
>
> Assume throughout that a 1 cm optical cell is in use.

We turn now from proteins to nucleic acids. All purine and pyrimidine bases absorb strongly in the 240–280 nm region, so that all the building blocks are chromophores, not just a small proportion of them as in proteins.

> **ITQ 13** Which would you expect to give the highest molar absorbance, nucleic acid or protein? (Let the nucleic acid be a tRNA which has approximately the same molecular weight as the protein.)

We have now introduced two characteristics of UV absorption—the wavelength at which an absorption peak occurs (λ_{max}) and the molar absorbance (ϵ) at that wavelength. Both these characteristics are very sensitive to the molecular environment of the chromophores concerned. This of course will change whenever the macromolecule alters conformation, and we can now turn to the use of absorption spectroscopy in conformational studies.

2.6.2 Conformational changes around individual amino acid residues

Because relatively few of the amino acid sidechains in a protein absorb in the UV, absorption spectroscopy can be used to study individual residues. Here we shall concentrate on tyrosine, although much of what follows applies equally well to tryptophan or to artificial chromophores introduced by chemical modification.*

In working with whole proteins rather than isolated amino acids, the first problem is deciding where in the absorption spectrum to look for specific changes.

> QUESTION Tyrosine in the active site of the enzyme carboxypeptidase undergoes a marked conformational change during substrate binding, as described in Section 2.2.3. Given the absorption spectrum of a solution of pure tyrosine (Figure 31), at what wavelength would you expect the absorption of whole protein to change most on substrate binding?
>
> ANSWER 280 nm. This is the λ_{max} for a solution of pure tyrosine.

In suggesting this wavelength, we have overlooked two problems in assigning peaks to individual amino acid residues. One is that the absorption characteristics of tyrosine may not be the same in free solution as in the microenvironment of the protein, and the other is that even a medium-sized protein contains several chromophores and their absorption peaks may very well overlap.

One way round this assignment problem is to give up trying to associate each part of the spectrum with a specific residue and to concentrate on particular peaks which *change* under different sets of conditions. This is the technique of *difference spectroscopy*. It side-tracks the problem by calculating the difference in absorption by the same protein under two sets of conditions. Only those parts that change show up, and these can be seen as peaks or troughs in the difference spectrum. Very often no calculation is necessary as the two spectra can be compared directly, making use of some device such as the split beam spectrophotometer shown in Figure 30. For difference spectroscopy the blank would contain not solvent but the macromolecule under the second set of conditions.

Figure 32 gives an example of difference spectroscopy being used to demonstrate what has long been suspected about enzyme conformation—that in many cases it changes on binding of substrate. Here the enzyme is TIM and a substrate analogue rather than true substrate has been used.

> QUESTION Substrate-induced conformational changes in TIM, as shown in Figure 32, may be produced under a series of different conditions (of temperature, pH, etc.) In order to compare these, it is experimentally simpler to make observations at a single wavelength rather than to scan over the whole wavelength range. Use the data in Figure 32 to select a suitable wavelength for this.
>
> ANSWER 290 nm. This is the position of the largest trough (or peak), and is the wavelength at which the spectrum is most sensitive to conformational changes in the enzyme.

A similar example of the role of spectroscopy in elucidating principles of molecular biology is given under Chemical modification (Section 2.9).

Spectrophotometric titration

In the living cell the precise pK** values of the ionizable residues of macromolecules are extremely important. They give an idea of the state of ionization of each residue and this, as we pointed out in Unit 1, will influence both the tertiary structure of the macromolecule itself (through changes in ionic bonding) and its interaction with other molecules in the cell. Unfortunately, the pK value of an individual residue in a macromolecule cannot be calculated from its value in free solution because of the influence of nearby residues. One way of measuring this important characteristic for a residue *in situ* within the macromolecule

assignment problem

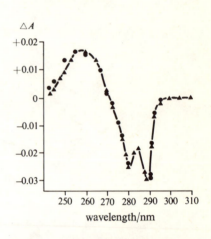

Figure 32 Substrate-induced conformational changes in TIM, as demonstrated by difference spectroscopy. The experimental set-up was similar to that of Figure 30. Each cell contained samples of the same enzyme solution, one of which was mixed with a substrate analogue.

---

* Chemical modification is another way round the assignment problem. Here the chromophore of interest is chemically altered in some way that affects its absorption characteristics. We shall return to this approach in Section 2.9.

** pK value is discussed in the *Source Book*. Briefly, it is the pH at which a residue is 50 per cent ionized.

is by *spectrophotometric titration*—a technique in which absorbance of a given chromophore is plotted against pH. The result is a titration curve which shows how absorbance changes with pH. From this can be calculated the parameter we are interested in, namely the pK of the ionizing chromophore. This approach can be applied to several branches of spectroscopy, but here we shall concentrate on one, UV absorption. We also consider only one residue, namely tyrosine. The reason why pH has such an effect on the UV absorption of this residue is that its ionized form has rather different absorption characteristics from its un-ionized form. (Figure 33 shows both the ionization equation and the UV absorption spectrum of the two forms.)

spectrophotometric titration

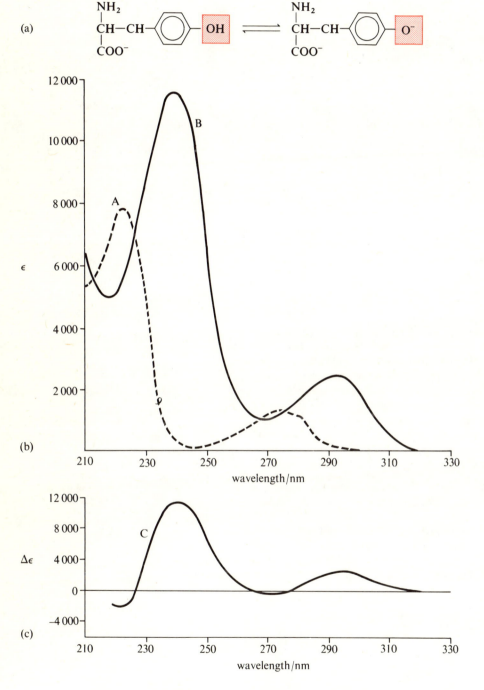

Figure 33 UV absorption spectrum of tyrosine, before and after ionization. (a) Equation for ionization; (b) spectrum (A) before ionization, at pH 6, and (B) after ionization, at pH 13; (c) difference spectrum, C, comparing tyrosine before and after ionization (note that the scale is now $\Delta\epsilon$, not ϵ).

QUESTION 1 What can you say from Figure 33 about the changes in (i) λ_{max} and (ii) absorbance, following ionization of tyrosine?

2 At what wavelength would you set a spectrophotometer in order to follow ionization with maximum sensitivity? (Ignore changes below 240 nm. In most proteins, as opposed to the single amino acid shown here, this region of the spectrum is swamped out by absorption from the peptide bonds, and is therefore difficult to interpret.)

ANSWER 1 (i) The λ_{max} has shifted from 278 to 295 nm. (ii) Absorbance at the old λ_{max} has fallen slightly, while that at the new λ_{max} has increased nearly two-fold. The net change at the new λ_{max} is even more startling when expressed in curve C as a difference spectrum, calculated by subtracting curve A from curve B. (Although the scale is expressed in terms of molar absorbance ϵ rather than absorbance A, the two are directly related, ϵ referring to the absorbance of a 1 M solution in a 1 cm cell, and A referring to any given solution.)

(2) The spectrophotometer should be set at the position of the highest useable peak in the difference spectrum, namely 295 nm.

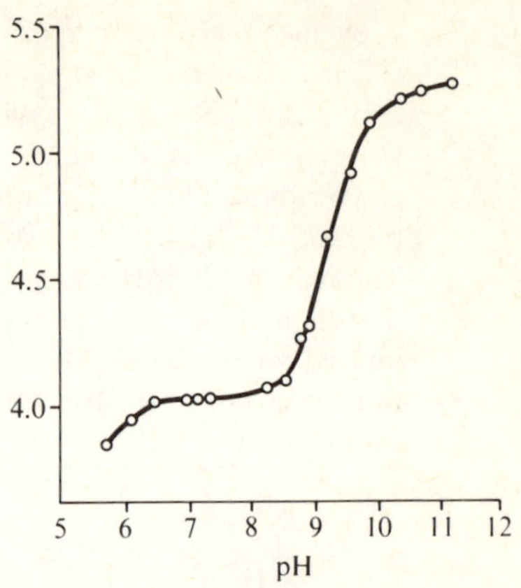

If the absorption at 295 nm (corresponding to the proportion of fully ionized form) is plotted against pH, a typical titration curve results (see Figure 34). The midpoint of this corresponds to the pH for 50 per cent ionization—which is one definition for the pK of an ionizable group. Hence this important parameter can be determined directly from the titration curve.

If we consider a whole protein rather than an isolated amino acid, many other ionizable residues will also be titrating in this pH range. Here we come to one of the big advantages of a spectrophotometric titration. Since tyrosine is the only residue contributing to the UV spectrum around 295 nm, we have been able to single it out for observation. In this way spectrophotometric titration can give the pK of individual residues within a macromolecule.*

Figure 34 Spectrophotometric titration of a hormone-binding protein that contains only one tyrosine residue. The spectrophotometer is set at the λ_{max} for fully ionized tyrosine, so that each reading gives a measure of the percentage of tyrosine in the ionized form. The midpoint of the curve gives the pK value for tyrosine in this protein.

2.6.3 Investigating secondary structure in nucleic acids

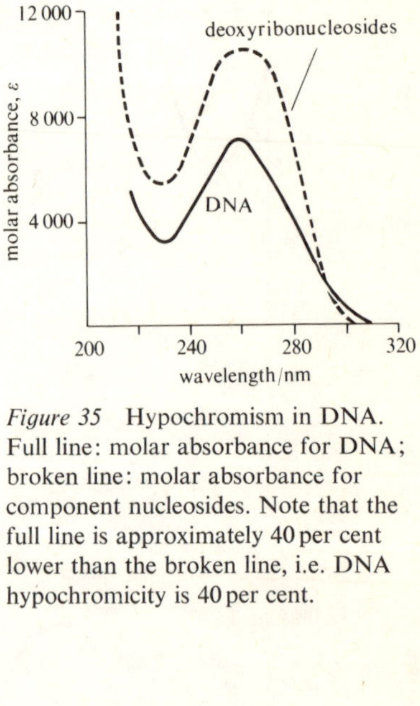

Nucleic acids do not lend themselves to the study of small conformational changes like those we have just described for proteins. This is because all their residues, rather than just a small proportion of them, absorb in the UV, and it is impossible to pinpoint small changes around individual chromophores. Conformational studies on nucleic acids therefore centre on the larger conformational changes that result from the disturbance of the regular secondary structures we described earlier. In these, the base chromophores lie stacked parallel to one another, and can no longer interact independently with incoming radiation. This results in *hypochromism*, a phenomenon whereby the macromolecule as a whole absorbs less strongly than would be expected from summing the individual absorbances of its constituent nucleosides.** You can see this effect in Figure 35 where the absorbance of DNA has been compared with that of its components, assuming these to act independently. Hypochromism is very sensitive to conformation, since the interaction between bases will change whenever one base alters its orientation relative to another. It can therefore be used to estimate the degree of regular secondary structure in a molecule.

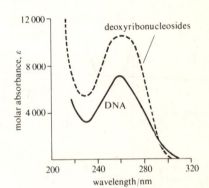

Figure 35 Hypochromism in DNA. Full line: molar absorbance for DNA; broken line: molar absorbance for component nucleosides. Note that the full line is approximately 40 per cent lower than the broken line, i.e. DNA hypochromicity is 40 per cent.

QUESTION What would you expect to happen to the UV absorption of single stranded helical RNA on heat denaturation?

ANSWER UV absorption will increase. This is because heat denaturation leads to loss of secondary structure, i.e. loss of the base stacking which allowed the chromophores to interact in a concerted fashion. There is no hypochromism in a denatured random coil molecule, therefore the observed absorption rises.

You might have guessed this answer from Figure 18, where we showed how UV absorption is used to demonstrate cooperativity during denaturation. In this case the nucleic acid was double stranded, and this introduces another factor into the picture, the difference between double and single stranded molecules. In the question just cited we deliberately chose a single stranded molecule because this is not stabilized by hydrogen bonding between complementary base pairs, but

---

* You will come across this technique again in the discussion on ribonuclease (Unit 4), where NMR rather than UV absorption is the spectroscopic method of choice.

** Nucleoside† is the base–sugar component of a nucleotide (base–sugar–phosphate). The base, of course, rather than sugar or phosphate, is what absorbs around 260 nm.

only by base stacking. In fact, H bonding seems to promote chromophore interaction, and hypochromism is considerably more pronounced for double stranded than single stranded nucleic acids. It is 40 per cent in DNA, compared with 15–20 per cent in single stranded RNA. (These figures are calculated by taking the sum of the theoretical absorbances of the individual constituent nucleosides as 100 per cent and comparing this with the actual absorbance of the nucleic acid.)

This difference in hypochromism has been put to good use as evidence in support of the cloverleaf structure of tRNA. In this model there are double stranded regions along the stems of each loop and single stranded regions in the bulges at the ends. If the hypochromism observed experimentally for tRNA is compared with that of model compounds (poly C for single stranded molecules and poly A/poly C* for double stranded), one can calculate that 50 per cent of the molecule would be expected to be double stranded. This is entirely consistent with the cloverleaf model, as you can estimate for yourself by looking again at Figure 21.

2.7   Other spectroscopic techniques

Study comment   Of the three techniques given here, only NMR is described in any detail. The physical principles of fluorescence, and of ORD/CD, are given briefly in Appendices 1 and 2, which are optional reading. That part of Section 2.7.3 that relates to the physical principles of NMR is not needed for fulfilling the Objectives of this Unit, but will be necessary for a full understanding of Unit 4.

2.7.1   Fluorescence

Much of what we have just said about UV absorption applies to other spectroscopic techniques which you may meet. Each of these has its own particular advantage. Fluorescence, for example, is far more sensitive than absorption spectroscopy, needing some 40-fold less of the macromolecular material. It is a popular method for studying the conformation of lipoprotein in membranes and its scope may be further increased by artificially introduced fluorescent probes.** These add to the information available from native fluorescing chromophores —tyrosine, tryptophan and occasionally phenylalanine. You will recognize that these are the three UV-absorbing residues of proteins. All five nucleic acid bases also fluoresce. The main difference between absorption and fluorescence spectroscopy is that in the latter an *emission* rather than an absorption spectrum is measured. The absorbed radiation is re-emitted in a manner characteristic of a particular chromophore in a particular molecular environment.‡ It is this effect of environment on fluorescence emission which makes it so useful for following changes in conformation. Fluorescence spectroscopy is used, particularly in proteins, to monitor changes around individual chromophores.

emission spectrum

IF YOU HAVE TIME, YOU SHOULD NOW READ APPENDIX 1.

2.7.2   Circular dichroism (CD) and optical rotatory dispersion (ORD)

Both circular dichroism and the closely related optical rotatory dispersion are often used to follow widespread changes in secondary as well as tertiary structure. This is because they are particularly sensitive to asymmetry‡‡ in macro-

---

* Poly A and poly C denote polymers of adenosine and cytidine, respectively.

** Probes or reporter groups are artificial chromophoric groups introduced into a macromolecule by chemical modification.

‡ More details of the physical principles behind fluorescence are given in Appendix 1, which is optional reading.

‡‡ An asymmetric molecule is one which lacks any plane of symmetry. For small organic molecules, a rapid way of detecting this is to look for an asymmetric carbon atom, i.e. one with four different substituents. However, even if it has no such asymmetric carbon a larger molecule may still lack a plane of symmetry due to its overall shape.

molecules. Repeat units like the amino acids may themselves be asymmetric, but it is their arrangement within the macromolecule that is more important. Pasteur likened the asymmetry of repeat units to that of the individual steps in a staircase, and the asymmetry of the whole molecule to that of a spiral staircase built from these blocks. Both the DNA duplex and the protein α-helix are regular secondary structures, comparable to the spiral staircase. Any slight change in the relative orientation of their component bases or amino acids will have far-reaching repercussions for the asymmetry of the molecule as a whole.

Methods for detecting this molecular asymmetry are given in Appendix 2, which is optional reading. The spectra can be used to calculate the amount of double-strandedness in a molecule or to follow cooperative conformational changes during denaturation. You have already come across this second application in the heat denaturation of DNA in Figure 18. Instead of A_{260}, one could here use ORD or CD to follow the conformational change and arrive at the same value for T_m. Furthermore, CD can detect changes over a greater temperature range than UV absorption.

2.7.3 Nuclear magnetic resonance (NMR)

Until recently, this technique was restricted to structural work on small molecules and it was not possible to distinguish anything useful from the spectra of macromolecules. Now that high resolution equipment is available, magnetic resonance spectroscopy, particularly that of the proton (i.e. the hydrogen nucleus), has excited a great deal of attention from biochemists. It can be applied to any atom whose nucleus has a non-zero spin quantum number.[A] This is a mathematical function describing those energy levels within a molecule that result from its magnetic properties. The commonest atomic nucleus* observed in NMR work is that of hydrogen, but others may be introduced to widen the scope of the method.**

The non-zero spin quantum number of these nuclei effectively means that they respond to a magnetic field as if they were tiny bar magnets. The bar has to take up a specific orientation relative to the applied field and only two such orientations are allowed. Continuing our bar magnet analogy, we may take these orientations as being either along the line of the magnetic field or opposite to it. The important point is that these two allowed states differ in energy. Therefore transitions between them (i.e. changes in orientation of the nuclear magnet relative to the applied field) will be accompanied by the uptake or emission of electromagnetic radiation. This allows the transitions to be studied by spectroscopy just like other transitions between energy levels[A] (e.g. electronic, vibrational). Two differences between NMR and other branches of spectroscopy are important. First, the energy gap between the two allowed states is very small, so that radiation of low energy (i.e. low frequency[A] or long wavelength) in the radio-frequency range must be used to study the transitions. Second, the energy gap appears only when a magnetic field is applied and is proportional to the magnetic field experienced by the nuclei. The precise field 'experienced' will depend mainly on the field applied to the sample by the experimenter, but also partly on the molecular environment of the spinning nucleus. This last, of course, is why biochemists use NMR to follow conformational changes in macromolecules.

What NMR measures, therefore, is the energy absorbed from a beam of radiation (in this case high frequency radio waves) of various frequencies. (How this is done is not vital to our discussion but since the resonance frequency of a particular nucleus is proportional to the applied field, for technical reasons the frequency is often held constant and the magnetic field varied. A simple calculation converts the resulting spectrum of absorbance against field strength, to one of absorbance against frequency.) As you can see in Figure 36, the NMR spectrum is similar to the UV spectra previously described, except that the positions of

---

* Remember that an atom is composed of a nucleus surrounded by an electron cloud.

** Isotopes of fluorine and phosphorus with non-zero spins have also been extensively used for NMR, but mainly in the form of probes which are introduced into specific areas of the macromolecule just like other spectroscopic probes of conformation. Another nucleus with non-zero spin is ^{13}C, and there is hope that this may be incorporated into the protein in place of ^{12}C (the 'normal' isotope) to give high resolution structural data of the type anticipated for proton NMR.

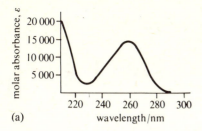

(a) wavelength/nm

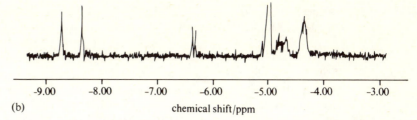

(b) chemical shift/ppm

the absorbance maxima are measured not as wavelengths or even frequencies but as the rather odd quantity *chemical shift*. Chemical shift is a measure of the amount by which the particular microenvironment of a nucleus causes the position of its resonance maximum to differ from that of a *reference nucleus*. In proton magnetic resonance, where H is the nucleus observed, the three common reference protons are those in tetramethyl silane (TMS, $(CH_3)_4Si$), hexamethyl silane (HMS, $(CH_3)_3Si-Si(CH_3)_3$) and chloroform ($CHCl_3$). A small amount of one of these compounds is included in each NMR sample to establish the reference frequency. Chemical shifts may be expressed in cycles per second (Hz), giving the shift in frequency from the reference point in a particular magnetic field, e.g. Hz from HMS. Nowadays they are more often expressed in parts per million (ppm), giving the *ratio* of the shift in frequency to the reference frequency, e.g. ppm from TMS.

There are, of course, plenty of protons around in macromolecules, so that in theory it is possible to observe every atom in the molecule simultaneously, giving a picture in solution comparable in detail to that obtained by X-ray crystallography in the solid state. This brings out the great advantage of proton NMR, and explains the initial excitement about this technique. Unfortunately, just because so many residues contribute to proton NMR, the assignment problem is a great deal worse than in any other spectroscopic method. So far it has been overcome for only a few special cases. A further drawback is that NMR requires rather large quantities of purified macromolecules, and for technical reasons these molecules should be heat stable to at least 40 °C.

Despite this, there is hope that the assignment problem will gradually be overcome, and meanwhile NMR continues to be used for high resolution work on small molecules bound to enzyme active sites, etc., and for following changes in the secondary structures of nucleic acids and polysaccharides.

Figure 36 NMR and UV spectra of adenine derivatives. (a) UV absorption of adenosine. (b) NMR spectrum of 5′-adenylic acid. Peaks in (b) represent points of maximum *absorbance of resonance energy.*

chemical shift

2.8 X-ray diffraction

Study comment We have gone into this very high resolution method in some detail because of its importance in the history of molecular biology. The TV programmes on lysozyme and on polysaccharides both illustrate different aspects of it—X-ray crystallography for lysozyme and fibre diagrams for polysaccharides—and should aid your understanding of the relevant Sections both here and in Unit 3. Stereoslide 10 is needed at the very end of the Section.

In this Unit we concentrate particularly on diffraction from crystals, but you should be able to explain the difference between this and diffraction by fibres in terms of (a) regularity of macromolecules in the preparation and (b) detail of information obtained. You should also be able to relate the practical difficulties of obtaining an electron density map, starting from a pure preparation of macromolecule, and to explain why a low resolution map is calculated from the central rather than the peripheral spots in a diffraction diagram.

None of the spectroscopic methods, whether they involve rotation of incoming radiation (as in ORD) or absorption (as in UV absorption and fluorescence), comes anywhere near X-ray diffraction in the matter of resolving power. X-rays have a wavelength of around 0.1 nm, which is much shorter than that of even UV radiation. More important, it is of the same order of magnitude as the distance between the atoms themselves. Therefore, in theory, it is possible to discriminate even between individual atoms in the macromolecule. Because of its high resolving power, X-ray diffraction has made perhaps more impact than any other method on our understanding of molecular biology. It provided the critical experimental evidence for Watson and Crick's famous model of the

DNA double helix, published in 1953. In 1961 it enabled the first high resolution map of a globular protein to be produced. Since then, many other proteins and a nucleic acid, tRNA, have been analysed at high resolution and have given the kind of detailed knowledge from which the principles of conformation enumerated earlier were built up.

In this account we shall not attempt to explain the physical background of the technique. We hope, however, to introduce enough terminology for you to understand the results of X-ray diffraction when described in non-technical scientific papers (non-technical here referring to the X-ray technique itself, not to the biochemical interpretation of the results!), and you should be able to appreciate the difference between X-ray diffraction by fibres and by crystals—the type of molecule and the level of detail obtained being different in each.

X-ray diffraction contrasts with spectroscopy in that there is little absorption of the incident radiation. Instead, we are interested in the *scattering* which occurs when this radiation comes into contact with particles or molecules within the preparation. If these molecules have a regular pattern in their orientation and position, the scattered rays can reinforce one another, adding up to give *diffraction maxima*. These are what produce the 'spots' in a diffraction diagram like those in Figures 37 and 38. Randomly oriented molecules, on the other hand, do not allow the necessary reinforcement of scattered rays and no diffraction pattern is seen.

diffraction maxima

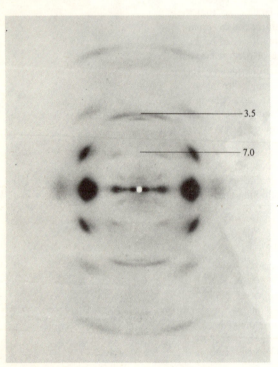

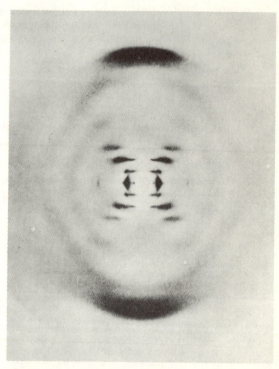

Figure 37 Fibre diagram of silk. *Note* Numbers refer to ångström spacing, not layer lines.

Figure 38 Fibre diagram of DNA.

The information that can be derived from diffraction patterns depends upon the degree of regularity within the macromolecular preparation, and this is where fibres and crystals differ. We shall begin with fibres, which are the simpler of the two. Fibrous proteins, DNA and many polysaccharides can be drawn out into threads or fibres, where the molecules are aligned with sufficient regularity along the fibre axis to produce a diffraction pattern known as a *fibre diagram*. With a comparatively simple fibre diagram, these diffraction maxima are recorded directly as spots on a photographic plate.

fibre diagram

QUESTION   Why does photography come into this?

ANSWER   Because a diffraction maximum comes from scattered X-rays, and X-rays can blacken a photographic plate.

The position of diffraction maxima in the diagram gives information on the *distance between repeat units* in the macromolecule itself. Take the fibre diagram of silk (Figure 37). The 7 Å *spacing* means that the diffraction maximum is produced by reinforced scattering from repeat units which are 7 Å apart in the molecule. The amino acid repeat distance in silk, for instance, is at 3.5 Å, and this gives a strong spot which you can readily see in the fibre diagram. The two amino acid repeat distance is also prominent at 7 Å—but this, for reasons of reciprocal space which we shall go into in a moment, occurs nearer the centre, not further away as you might expect. Silk has a β-structure and there is nothing particularly remarkable about its diffraction pattern. A helix, on the other hand, is immediately recognizable from the characteristic *helix cross* clearly visible in the diffraction pattern of DNA (Figure 38). It was just such a picture that Rosalind Franklin (with R. G. Gosling and M. H. F. Wilkins) produced in the early 1950s, from which Watson and Crick (1953) postulated that DNA, the previously elusive genetic material, had a double helical structure.*

We come now to the high resolution work that is possible using crystals rather than fibres. The extra degree of three-dimensional order in a crystal gives rise to many more diffraction maxima, and the extra resolution comes from paying particular attention to those that land on the periphery of the diffraction diagram. To explain this we need to go into the concept of *reciprocal space*. Because of the way in which diffraction maxima are generated, scattering from points that are *far* apart in the crystal gives rise to diffraction maxima that are *close* to the centre in the diffraction diagram. These can be used to outline the molecule. Conversely, points that are close together in real space will give spots on the edge of the diffraction diagram, and these can be used to fill in the detail. Figure 39 should make this clearer. The numbers (expressed in ångström) refer to *resolution*, i.e. to the minimum distance between points (in real space) that can be distinguished from one another. As you can see, the lowest resolution picture (2.2 Å) is calculated solely from data in the innermost section of the reciprocal

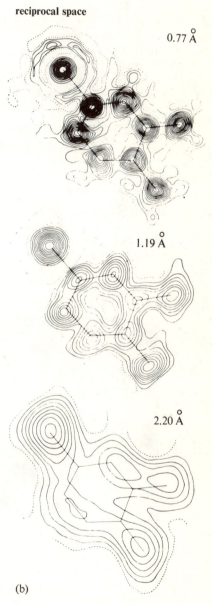

reciprocal space

0.77 Å

1.19 Å

2.20 Å

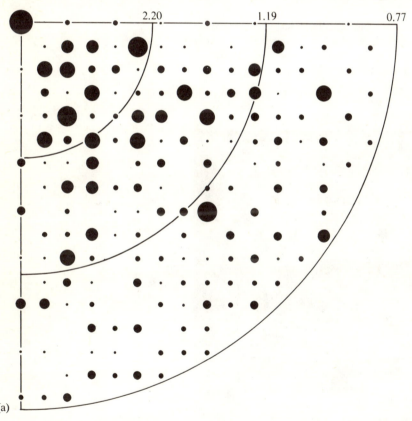

(a)

(b)

space diagram, and this gives only a vague outline of the molecule. The sharply defined high resolution picture uses all the data, including that from points as close together in real space as 0.77 Å.

Only if the low resolution map looks promising is it 'refined' to high resolution by including spots from further out. The reason for this caution is obvious when you consider the labour involved in measuring both the position and intensity of several thousand spots. Originally this would have been done by hand but nowadays the information is commonly fed straight into a computer.

Figure 39 Resolution and reciprocal space. (a) Part of the reciprocal space diffraction diagram from which maps were calculated. Each quarter-circle encloses the reciprocal space points used to build a map at the resolution indicated. Note that high resolution (0.77 Å) uses all the points, while low resolution uses only a fraction of them. (b) Two-dimensional electron density map of a heavy-atom pyrimidine derivative at 0.77, 1.19 and 2.20 Å resolution.

* Helical parameters n, p and d can be derived directly from a fibre diagram, as will be described in Unit 3 and in the TV programme on polysaccharides.

One of the major difficulties at the low resolution stage is the phase problem, and to appreciate this we need to consider what exactly it is we are trying to calculate. With a fibre diagram we were aiming only at estimating distances between repeat units in the molecule, but with crystallography it should be possible to calculate a complete three-dimensional electron density map. This can be computed from the experimental data available—namely the position and intensity of spots—provided one further piece of information is forthcoming, the *phase*. As you may remember from earlier Courses, all electromagnetic radiation[A] is composed of wave functions, and what the phase does is to fix where abouts on the wave the other two sets of experimental measurements have been taken—at a trough, a peak or midway between.

phase

Unfortunately, no method of X-ray detection is able to measure the phases directly, and at present the phase problem is best solved by *isomorphous replacement*. Most proteins are composed entirely of 'light' atoms, O, C, N, H and S, none of which are particularly outstanding in their ability to scatter X-rays. But if a heavy atom like a metal can be inserted into the molecule it will so dominate the diffraction pattern that it can act as a reference point from which the missing parameter—phase—can be unambiguously determined. In practice, crystals of several such heavy-atom derivatives are needed, and procuring them is one of the main problems in X-ray crystallography. (Another one is persuading the native macromolecule to crystallize in the first place!)

isomorphous replacement

By whatever means the two proteins (with and without heavy atom) are prepared, it is essential that they retain their identity in all respects other than the presence of this one atom. If there are any changes other than those associated with its insertion or removal, the two molecules can no longer be compared for phase calculation purposes.

Finally we come to the point where all problems seem to have been overcome. The protein has been persuaded to crystallize, suitable isomorphous derivatives have been prepared, the phase problem has been solved, and the intensity and position of diffraction maxima have been measured. What comes next?

THE ANSWER IS SHOWN IN STEREOSLIDE 10, WHICH YOU SHOULD NOW STUDY.

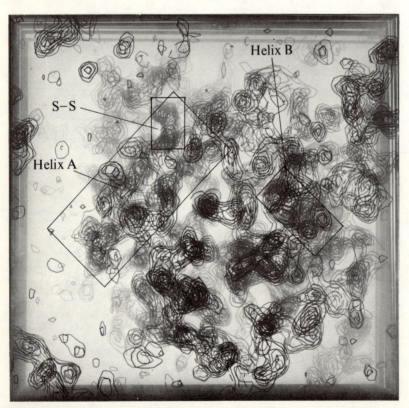

Figure 40 Identification of helices in Stereoslide 10, the electron density map of lysozyme at 2 Å resolution. The helix marked A (residues 5–15) runs parallel to the plane of the section, while the helix marked B (residues 24–35) runs vertically into the map, turning slightly to the right. A disulphide bridge (linking residues 6 and 127) can be seen rather faintly at the area marked S—S.

This shows part of an *electron density map*, which is perhaps the nearest we can ever come to visualizing what a protein really looks like. Here you see a series of horizontal sections through the protein, lysozyme. The contour lines in each section join points of similar electron density, just as in a geographical map they join points of similar height. By superimposing a series of such sections we can build up a three-dimensional picture of the molecule in depth. With the aid of Figure 40, you should be able to make out two stretches of α-helix in this region of the lysozyme map.

electron density map

It is from a map like this that the positions of individual atoms in the molecule have to be deduced. Fortunately, atoms come packaged in recognizable elements of structure. To pick these out from the map, one has to imagine the structural formula superimposed on the electron density. We showed this earlier, in Figure 22, for a nucleic acid. As you can see, interpreting electron density maps needs a feel for chemical formulae in terms of shape rather than just symbols, and you may not be surprised to learn that no protein or tRNA has yet been fully interpreted without prior knowledge of its sequence.

2.9 Chemical modification

We move now from physical techniques to chemical modification, a technique which can be used either as an aid to spectroscopy and X-ray diffraction—by inserting probes or heavy atoms into strategic places—or as a method on its own, to map the position of individual residues in the tertiary structure of the molecule. You should be able to give examples of these probes and also interpret results of mapping experiments.

In both applications a major problem in chemical modification is to achieve specificity of reaction, i.e. to make sure that only the residue under investigation has been modified. This is discussed in Section 2.9.3.

2.9.1 Mapping residue positions in the tertiary structure

It is impossible to tell from primary structure data alone, just where in the final folding pattern a particular residue may lie. Spectroscopic techniques can offer little information on this point and X-ray crystallography is applicable only where the macromolecule will oblige by crystallizing. The principle behind using chemical modification for mapping is *conformational shielding*, whereby certain residues show suppressed chemical reactivity because of their position in the tertiary structure. This usually implies that they are buried in the interior, rather than exposed on the surface of the molecule.* An example should make this clearer. A protein known (from amino acid composition) to contain two tyrosines is reacted with a powerful nitrating agent such as tetranitromethane. If these tyrosines had behaved as in free solution you would expect both to be rapidly converted to 3-nitrotyrosine, as shown in equation 2.2.

conformational shielding

(2.2)

* In nucleic acids, which do not always have a unique tertiary structure, the term is also applied to suppressed reactivity caused by H bonding in the secondary structure, e.g. residues in the stem rather than the loop of a tRNA arm will show conformational shielding and will be less reactive.

In the intact protein, however, the situation can be quite different, as shown in Table 6.

TABLE 6 Mapping the position of tyrosine residues by chemical modification

Protein solution used for reaction with TNM	Amino acid composition of protein *before* reaction		Amino acid composition of protein *after* reaction	
	mol Tyr/mol protein	mol 3-NT/mol protein	mol Tyr/mol protein	mol 3-NT/mol protein
1 Protein in buffer	2	0	1	1
2 Protein in urea	2	0	0	2
3 Protein in buffer plus substrate	2	0	2	0

TNM = tetranitromethane; 3-NT = 3-nitrotyrosine.

QUESTION   From lines 1 and 2 in Table 6, what can you conclude about the possible location of the two tyrosines in the tertiary structure of the protein? (Remember that urea is a denaturant.)

ANSWER   One tyrosine appears to be exposed on the surface, readily available for reaction, while the other seems to be inaccessible and is probably buried in the interior. The extra evidence in line 2 is needed to confirm that it is indeed the tertiary structure that is shielding the second tyrosine from reaction. When this structure is disrupted by denaturation in urea, both residues become freely available for nitration.

QUESTION   From the evidence in line 3 of Table 6, what further suggestions can you make about the whereabouts of the exposed tyrosine on the surface?

ANSWER   The exposed tyrosine may be located in the active site, since the presence of substrate appears to shield it from reaction. Note, however, that this last result should be interpreted with caution. Conformational changes induced by substrate binding can be quite far-reaching, and could result in indirect shielding of residues even remote from the active site. The evidence in line 3 is insufficient on its own to eliminate this second possibility.

Here you begin to see some of the limitations of the technique when applied to mapping residue positions. However, this is not the only reason why molecular biologists turn to chemistry, and we shall now move to the second role of chemical modification in conformation studies.

2.9.2 Introducing reporter groups

As you may have noticed in earlier Sections, some molecules seem rather perversely designed from the point of view of spectroscopy. They either lack suitable chromophores in interesting places, or else they have so many of one kind that it is impossible to distinguish between them. Chemical modification can sometimes improve on this situation by inserting probes or reporter groups into the molecule, or conversely by blocking some off when there are too many.* Here, however, we shall concentrate on the introduction of reporter groups, returning you to equation 2.2 where the nitro group introduced by tetranitromethane is a good example of a probe for use in absorption spectroscopy.

reporter groups

* There is a good example of the blocking-off role of chemical modification in ribonuclease, where it was used to distinguish between the four different histidines (see Unit 4).

ITQ 14 The nitro group inserted as a probe into tyrosine turns it yellow. To be precise, the λ_{max} of the 3-nitrotyrosine formed lies at around 400 nm. What advantage does this have over unmodified tyrosine as a reporter group?

Inserting fluorescent probes into purified macromolecules and into membranes is a popular way of extending the scope of fluorescence, a very sensitive technique that would otherwise be restricted to looking at conformational changes around tyrosines or tryptophans. As another example of how optical methods and chemical modification go together you have already had the heavy atom 'probes' of X-ray crystallography.

Figures 41 and 42 show how chemical modification can resolve one of the main difficulties in interpreting UV absorption spectra. This is the problem of assigning peaks in the spectrum to specific residues in the molecule. In Figure 41, curve A, you can see that the protein as a whole absorbs right across the UV wavelength range. Much of this absorption is due to a single chromophore, tyrosine. This can be seen from curve B of Figure 41, where the single tyrosine in the molecule has been chemically modified (while *in situ* in the macromolecule) to acetyltyrosine, an analogue with very low UV absorbance.

This particular protein—neurophysin—is not an enzyme but a hormone carrier protein concerned in the transport of the antidiuretic hormone vasopressin. It has a binding site specific for its hormone in just the same way as an enzyme has an active site specific for its substrate. What was not clear until recently was that carrier proteins may undergo hormone-induced conformational changes similar to the substrate-induced conformational changes familiar in enzymes. Figure 42 shows some of the evidence for this. Addition of hormone alters the protein absorption spectrum, indicating some change in the relative orientation of chromophores within the macromolecule, i.e. a conformational change. As you can see from these data, one of the residues involved in this molecular movement is tyrosine.

QUESTION What is the advantage of using nitrated protein for this experiment rather than the native form?

ANSWER The absorbance peak due to tyrosine shifts to the visible region of the spectrum, where it can readily be distinguished from the absorbance of other chromophores. It then becomes possible to predict more precisely which residues may be involved in the conformational change.

2.9.3 Achieving specificity in chemical modification

In the nitrotyrosine example of Table 6 we had an ideal set-up for chemical modification. The reagent was specific for tyrosines, and the protein had only two residues of the type to be modified, one of which was shielded by the tertiary structure of the protein. Furthermore, the chemical reaction was mild enough to be carried out under non-denaturing conditions, so that the buried tyrosine remained buried. The result was a protein specifically labelled at one site only. Unfortunately, most situations are a great deal more complicated than this, and all three difficulties may arise, i.e. lack of specificity in the protein (too many residues of the same type), lack of specificity in the reagent (too many different types of residue react), and protein unfolding under the conditions of the reaction.

The first two of these difficulties may be overcome by exploiting the tertiary structure of the protein itself. Remember that in a native macromolecule each residue has its own particular microenvironment, and some of these may be more conducive to reaction than others.

QUESTION Would you expect residues in the following positions to have enhanced or suppressed reactivity towards reagents used in chemical modification?
(a) A residue buried in the interior of a globular protein.
(b) A residue forming part of the active site of an enzyme.

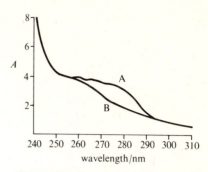

Figure 41 Effect of tyrosine modification on the UV absorption spectrum of a hormone-binding protein. Curve A: native protein; curve B: protein after conversion of tyrosine to acetyltyrosine.

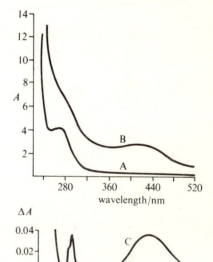

Figure 42 Difference spectroscopy used to follow hormone-induced conformational changes in a protein. The hormone *vasopressin* binds specifically to *neurophysin*, its carrier protein. Curve A: spectrum of native neurophysin; curve B: spectrum of nitrated neurophysin; curve C: difference spectrum comparing nitrated neurophysin in the presence and absence of bound hormone.

ANSWER   (a) A residue buried in the interior will have suppressed reactivity. This is a clear example of conformational shielding.
(b) Residues in the active site often show enhanced reactivity.

Active site residues may not always have this useful property of enhanced reactivity, but the active site as a whole will have enhanced *binding affinity* for any small molecule resembling the natural substrate. This property is exploited in a technique known as *affinity labelling*, whereby any small molecule for which the protein already has a specific binding site may be used to carry reactive reporter groups into particular areas of the molecule. This approach can be applied to any macromolecule with a binding site specific for one particular ligand (e.g. substrate, antigen, neurotransmitter), provided only that the ligand, in being altered to carry the reporter group, is not made unrecognizable to the macromolecule.

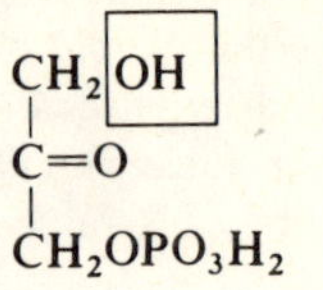

As an example of an affinity label designed for the active site of the enzyme TIM, compare compounds (a) and (b) in Figure 43. Dihydroxyacetone phosphate (DHAP) is a substrate for the enzyme and therefore binds strongly to the active site. This binding is not impaired if the C-1 hydroxyl is replaced by a bromine atom to give the compound you can see in (b). Bromine is a powerful nucleophile† and reacts rapidly with nearby groups in the active site. The enzyme is thereby irreversibly inhibited and the active site region labelled by a covalently bound substrate analogue.

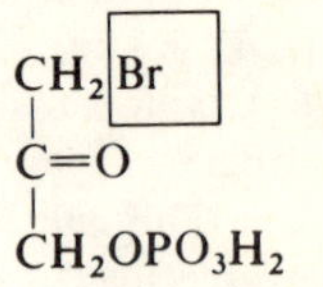

NOW THAT YOU HAVE COVERED ALL OF PART II YOU COULD TRY SAQS 11, 12 AND 13, WHICH RELATE TO THE CHOOSING OF EXPERIMENTAL METHOD, AND SAQ 14 WHICH CONCERNS CHEMICAL MODIFICATION ALONE.

$$CH_2\boxed{OH}$$
$$|$$
$$C=O$$
$$|$$
$$CH_2OPO_3H_2$$
(a)

$$CH_2\boxed{Br}$$
$$|$$
$$C=O$$
$$|$$
$$CH_2OPO_3H_2$$
(b)

Figure 43 Affinity labelling of the active site residues in TIM. (a) Dihydroxyacetone phosphate (substrate); (b) bromohydroxyacetone phosphate (substrate analogue with reactive Br group).

References

KIM, S. H., QUIGLEY, G. J., SUDDATH, F. L., MCPHERSON A., SNEDEN, D., KIM, J. J., WEINZIERL, J. and RICH, A. (1973) Three-dimensional structure of yeast phenylalanine transfer RNA: Folding of the polynucleotide chain. *Science, N.Y.*, **179**, 285–288.

LADNER, J. E., JACK, A., ROBERTUS, J. D., BROWN, R. S., RHODES, D., CLARK, B. F. C. and KLUG, A. (1975) *Proc. Nat. Acad. Sci. U.S.A.*, **72**, 4414–4418.

ROBERTUS, J. D., LADNER, J. E., FINCH, J. T., RHODES, D., BROWN, R. S., CLARK, B. F. C. and KLUG, A. (1974) Structure of yeast phenylalanine tRNA at 3 Å resolution. *Nature, Lond.*, **250**, 546–551.

WATSON, J. D. and CRICK, F. H. C. (1953) A structure for DNA. *Nature, Lond.*, **171**, 736–738.

Recommended reading

All the books and the comments on them in the Unit 1 Recommended reading list are relevant to this Unit, as well as the following:

BLOOMFIELD, V. A., CROTHERS, D. M. and TINOCO, I. (1974) *Physical Chemistry of Nucleic Acids*, Harper and Row.
A specialized account of the evidence for nucleic acid conformation.

BLOOMFIELD, V. A. and HARRINGTON, R. E. (eds) (1975) *Biophysical Chemistry: Readings from Scientific American*, Harper and Row.
Articles by Perutz, Kendrew, Phillips and Koshland are particularly relevant to this Unit.

DICKERSON, R. E. and GEIS, I. (1969) *Structure and Action of Proteins*, Harper and Row.
A highly readable, more specialized text.

Appendix 1 (Black)

Principles of fluorescence

When exposed to radiation at an appropriate wavelength (usually UV) many proteins are themselves capable of emitting radiation. This is usually fluorescence, the same phenomenon that makes detergent-washed white shirts glow under UV light.

A molecule that has absorbed radiant energy is said to be in an *excited state*,[A] in which some of its electrons have been raised from their ground state to a higher energy level.[A] When the electron returns to its ground state (S_0) this energy is dissipated in two stages (Figure 44). In the second of these, the transition $S_1 \rightarrow S_0$, one of three things may happen: energy may be emitted as light (i.e. fluorescence), it may be diverted into photochemical reactions (of great significance for photosynthesis, see Unit 8), or it may be *quenched* (see below) by transfer to other chromophores.

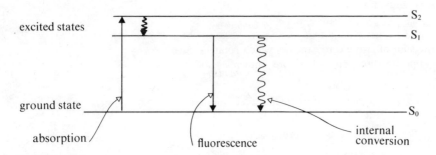

Figure 44 Energy level diagram to show the relationship between absorbed radiation and fluorescence emission.

The first of these possibilities, fluorescence emission, is shown by all five nucleic acid bases (A, G, U, T and C) and by the three aromatic amino acids. Each *emission spectrum* is characteristic of the chromophore in a particular environment.

QUESTION   The absorption and emission spectra for tyrosine are shown in Figure 45. What is the most striking point about λ_{max} for the two different spectra?

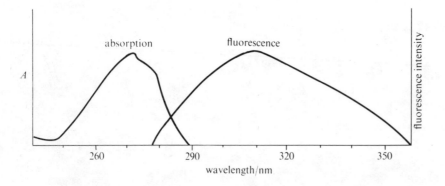

Figure 45 Absorption and fluorescence emission spectra of tyrosine.

ANSWER   The λ_{max} for emitted light is at a longer wavelength than for the absorbed light. Some energy is always lost in the conversion, and the resulting emitted radiation therefore has *lower* energy and a *longer* wavelength.

Fluorescence intensity is measured at λ_{max} in the emission spectrum of the macromolecule. For proteins this is usually near the λ_{max} for tryptophan, whose fluorescence tends to dominate the spectrum. When intensity at this wavelength is lower than expected, we have *quenching*. Most often this results from transfer of absorbed energy from one chromophore (the donor) to a second chromophore (the acceptor) which is not itself capable of emitting fluorescence energy.

54

As a result the fluorescence intensity drops sharply. The precise extent of this drop depends upon the efficiency of transfer between chromophores, this in turn depends upon their relative orientation, and this, as you might have anticipated, will be influenced by conformational changes in the molecule.

Ligand-induced conformational changes may be measured as a difference spectrum in the same way as for UV absorption, by comparing fluorescence intensity in the presence and absence of bound ligand.

Appendix 2 (Black)

Principles of optical rotatory dispersion (ORD) and circular dichroism (CD)

In order to detect molecular asymmetry by optical methods it is necessary to use polarized light.[A] You should already be familiar with the idea that electromagnetic radiation[A] can be represented by vectors[A] distributed randomly in the plane at right angles to the direction of propagation. This radiation can be polarized by devices which cut out all vibrations except those in a single plane (see, for example, the incident beam in Figure 46a). This plane polarized light can be represented formally by the sum of two electric vectors which rotate at equal angular velocities but in opposite directions—these are known as the left and right circularly polarized components, E_L and E_R (see Figure 46b).

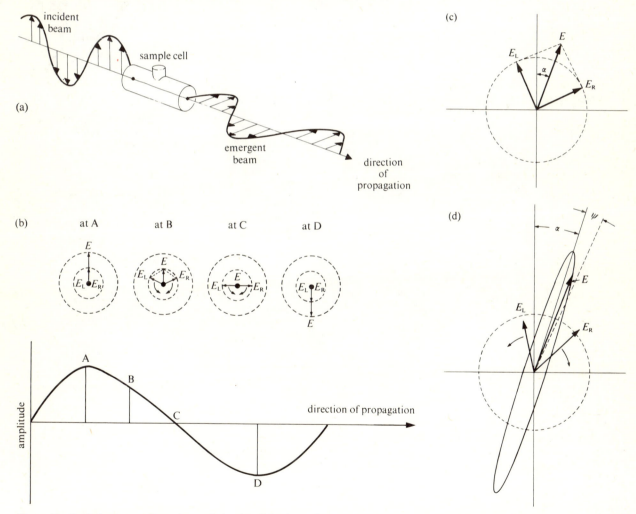

Figure 46 Optical rotatory dispersion and circular dichroism. (a) Plane polarized light before and after passing through a solution of asymmetric molecules (note that here, α is 90°). (b) Diagram to show how E, at each point along the wave, is the sum of vectors E_L and E_R. (Note that where components reinforce one another—as at A—the resultant E is at a maximum, whereas where they are equal and opposite—as at C—the resultant is zero.) (c) Unequal velocity gives ORD. (E_L and E_R are the same length, but have rotated through different angles.) (d) Unequal absorption—as well as unequal velocity—gives CD. (E_L and E_R have rotated through different angles, and are also different lengths.)

When polarized light passes through a symmetric substance these two components are similarly affected. Asymmetric molecules, however, will affect the components in different ways.

Where the two vectors rotate at different angular velocities in a solution of the asymmetric molecule, the resultant vector E will be at an angle to the original plane. This angle α—the degree of rotation of the plane of polarized light—is what is measured in ORD. In CD the two components not only rotate at different angular velocities, they are *absorbed* to different extents (i.e. one arrow—E_L or E_R—is shorter than the other). The result is that E no longer oscillates along a straight line but traces out an *ellipse*, as shown in Figure 46d. (This is sometimes known as elliptically polarized light.)

Older instruments measured this ellipticity, ψ, but modern instruments measure directly the difference in absorption ($A_L - A_R$).

Both the optical rotation and the differential absorption are dependent on the wavelength of the incident light. With both techniques, therefore, a spectrum is measured, showing the variation of the effect as a function of wavelength.

Most information can be gained by examining ORD or CD in the region of maximum absorbance. Hence proteins are examined around 220 nm, where the peptide bond absorbs; characteristic spectra are displayed by both α-helix and β-pleated sheet. Nucleic acids are investigated at around 260 nm in the range of absorbance of the nucleoside bases.

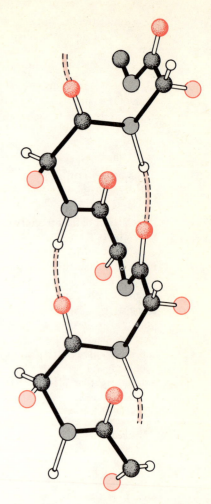

Figure 47 A 3_{10} protein helix. Atoms and H bonds are colour coded as in Figure 5. (For SAQ 1.)

Self-assessment questions

SAQ 1 (*Objective 1a*) Figure 47 shows a 3_{10}-helix. This is a type of secondary structure found in short stretches of lysozyme and similar to the α-helix shown in Figure 5 of this Unit. By comparing Figures 5 and 47, state:

(a) What is the value of n, the number of residues per turn of helix?

(b) If the repeat distance, d, is 2.0 Å, what is the pitch of the 3_{10} helix?

(c) How many stabilizing H bonds are there per turn of 3_{10} helix?

(d) Would you expect the H bonds in the 3_{10}-helix to be stronger or weaker than those stabilizing the α-helix?

SAQ 2 (*Objective 3a*) Are the following statements true or false?

(a) When an extended polypeptide chain folds up to its native conformation, it selects the one with maximum free energy.

(b) In an aqueous environment, the most stable conformations are the ones in which hydrophobic residues are inside and shielded, rather than exposed and outside.

Give a reason for your answer to (b).

SAQ 3 (*Objective 3a*) Table 7 shows the results of a denaturation–renaturation experiment in which catalytic activity was used to monitor unfolding and refolding of the protein chain. Would you expect the activity of solution C to be 75 EU or 2 EU,

(a) if the dialysis step, 3, took place immediately;

(b) if the protein were chemically modified after step 2 by acetylation, and then dialysed?

Hint The protein denaturant urea is a small molecule that can pass through a dialysis† membrane.

TABLE 7 Denaturation–renaturation data for SAQ 3

Protein solution	Treatment	Enzyme activity/EU*
A	1 Protein dissolved in buffer	100
B	2 Solution A above, made 6 M in urea	
C	3 Solution B above, dialysed exhaustively against fresh buffer	2

* Enzyme units, EU, are units of activity.

SAQ 4 (*Objective 2*) Carboxypeptidase (Stereoslide 1 or 2) has a wide stretch of β-sheet running vertically down the centre. (Look for the red H bonds running *vertically*, between adjacent chains.) How many different sections of polypeptide chain are linked together in this β-structure? Note that a β-sheet is composed of many horizontal chains stacked vertically.

SAQ 5 (*Objective 2*) About one-third of the way down on the left-hand side of this view of carboxypeptidase (Stereoslide 1 or 2) is a length of α-helix running almost exactly parallel to the horizontal edge of the slide.

(a) Approximately how many turns does this helix have?

(b) Is the total number of amino acids in this helix nearer to 10, 12, 16 or 25? *Hint* The numbers can be calculated rather than counted, from the number of residues per turn of α-helix.

SAQ 6 (*Objective 2*) What is the secondary structure of the section of protein chain indicated in lilac in the lysozyme sausage model shown on the right-hand side of Stereoslide 7? Use the other half of Slide 7, or Slide 6, to answer this question.

SAQ 7 (*Objective 3b*) Two different forms, X and Y, of a polysaccharide-hydrolysing enzyme can be isolated from a culture of bacteria. Form Y, the minor component, has only 5 per cent of the catalytic activity of form X. When examined in the ultracentrifuge, the respective sedimentation coefficients were found to be 56S (form X) and 34S (form Y). Which of the following statements gives the best interpretation of these facts? (Proteolytic degradation during isolation of the enzyme can be ruled out, since protease inhibitors were added as a precaution.)

(a) Enzyme Y is a mutated form of enzyme X in which a change has occurred in the amino acid sequence in the region of the *active* site.

(b) Enzyme X is a multimeric protein which loses most of its catalytic activity on dissociation. The mutated form, Y, has a change in amino acid sequence in the region of the subunit interaction site.

SAQ 8 (*Objective 1a*) Melting curves for DNA purified from two different sources—one animal and the other plant—gave T_m values of 40 and 86 °C respectively. Which DNA has the higher content of G+C residues?

SAQ 9 (*Objective 1a*) It has been suggested that heat denaturation of DNA takes place in two stages, as depicted in Figure 48. In the intermediate stage, (b), the small loops represent readily denatured regions where H bonding has already broken down. The straight portions represent helical regions still held together by H bonding between complementary base pairs. This pattern is thought to reflect differences in nucleotide composition. Would you expect the loops to contain mainly adenine–thymine or guanine–cytosine base pairs?

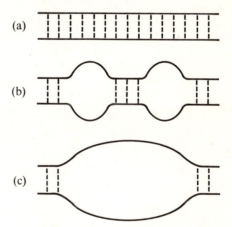

Figure 48 Loss of secondary structure during heat denaturation of DNA. (a) Native DNA; (b) partly denatured DNA; (c) almost fully denatured DNA.

SAQ 10 (*Objective 1a, 1b*) The following sequence shows the part of the primary structure of an imaginary tRNA$_{asp}$ that is thought to include the anticodon loop:

G-C-U-U-G-A-A-C-U-A-G-U-C-A-A-G-C

Suggest how this sequence may be folded in the secondary structure, remembering that maximum binding energy results when complementary base pairs are opposite one another (as in ITQ 10) and that the anticodon loop should be exposed. *Hint* The codon for aspartic acid, to which the anticodon sequence must be complementary, is GAU.

SAQ 11 (*Objective 4*) Which of the experimental techniques listed in Table 8 would you select for studying unfolding and refolding phenomena in the following macromolecular preparations? (Since the best studies employ a battery of techniques you may choose more than one method, if appropriate.)

(a) A protein lacking in both tyrosine and tryptophan residues,

(b) a DNA,

(c) a protein available only in small quantities,

(d) a tRNA.

TABLE 8 Experimental techniques for problems described in SAQs 11–13

ORD

UV absorption spectroscopy

X-ray crystallography

NMR

Mapping by chemical modification

Fluorescence emission spectroscopy

SAQ 12 (*Objective 4*) Which of the experimental techniques listed in Table 8 would you select for studying conformational changes in a non-crystallizable enzyme on binding of its substrate? (More than one technique may be chosen.)

SAQ 13 (*Objective 4*) Which of the experimental techniques listed in Table 8 would you select for a detailed study of tertiary structure in the following macromolecules? (More than one technique may be chosen.)

(a) An enzyme that will not crystallize, (c) a purified mRNA,

(b) haemoglobin, (d) a tRNA.

SAQ 14 (*Objective 5*) The following observations (a–c) were made on the enzyme carboxypeptidase in the presence and absence of phenylpropionate—a substrate analogue which binds in its active site pocket but is not hydrolysed by the enzyme.

(a) The UV absorbance at 298 nm of a carboxypeptidase solution changes upon binding phenylpropionate.

(b) A solution of nitrocarboxypeptidase (a modified form of the enzyme in which exposed tyrosine residues are converted by chemical modification into nitrotyrosine residues) shows a change in absorbance in the yellow region of the spectrum when phenylpropionate is added to the solution.

(c) After iodination, a tetrapeptide containing one iodotyrosine residue could be isolated from a partial hydrolysate of the protein. When iodination was repeated in the presence of phenylpropionate, the same tetrapeptide when isolated contained unmodified tyrosine.

Treating each observation as a separate experiment, which of the following conclusions (i or ii) can be drawn from the data in observations (a) to (c)?

(i) Tyrosine residues lie in the active site pocket and are protected from chemical modification by bound substrate.

(ii) When substrate binds to the active site the protein undergoes a conformational change somewhere in the region of tyrosine or tryptophan residues.

ITQ answers

ITQ 1 In Figure 49 the molecular orbitals around the rigidly held O, N and C atoms of Figure 2 are sketched in. This should make it clear that there will be no rotation about bonds N_A—C_A or N_B—C_B. Rotation about the other bonds,

$$C_A-\alpha C_B, \quad \alpha C_B-N_B \quad \text{and} \quad \alpha C_C-N_C,$$

does take place, as indicated by the arrows.

ITQ 2 The first conformation ($\phi = 40°$, $\psi = 315°$) describes a β-sheet while the second conformation ($\phi = 130°$, $\psi = 120°$) describes an α-helix. (Model building exercise 2 gives more details on the β-sheet.)

ITQ 3 Two. In this short section of β-structure there is a —C=O and an —N—H projecting on either side of the chain. (Figure 6 shows how these form H bonds between adjacent antiparallel chains.)

ITQ 4 They will be strong H bonds, because all three combining atoms (N—H—O) lie in approximately the same straight line.

ITQ 5 The R groups, which point vertically up and down. In silk, R is often CH_3 (from the amino acid alanine) and forms hydrophobic bonds between vertically stacked sheets.

ITQ 6 The first side, so that the non-polar residues alanine, glycine, phenylalanine and tryptophan can form van der Waals bonds with similar residues in the hydrophobic core.

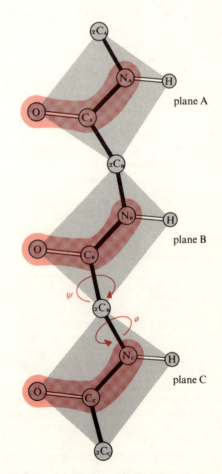

Figure 49 Tripeptide, showing flexibility in the relative orientation of planar peptide bonds.

This also has the advantage of leaving the polar residues (glutamic acid, lysine and aspartic acid) on the second side free to form H bonds with water molecules in the external medium.

ITQ 7 Hydrogen bonding. This is normally provided by the tyrosine OH. The $\alpha_1\beta_1$ contact is normally maintained in part by an H bond between the tyrosine OH on the β subunit and aspartic acid on the α subunit. This is a clear example of the way clinical symptoms at the whole organism level may be traced to defects at the molecular level.

ITQ 8 The number of H bonds per base pair is three for G+C and only two for A+T. Therefore the net binding energy increases as the ratio (G+C):(A+T). At high ratios the molecule is more resistant to denaturation, and therefore T_m is high.

ITQ 9 Double stranded DNA has pronounced plateau regions and a very much steeper slope to its melting curve than does single stranded DNA.

ITQ 10 (a) The hairpin loop with maximum base pairing is as shown in Figure 50.

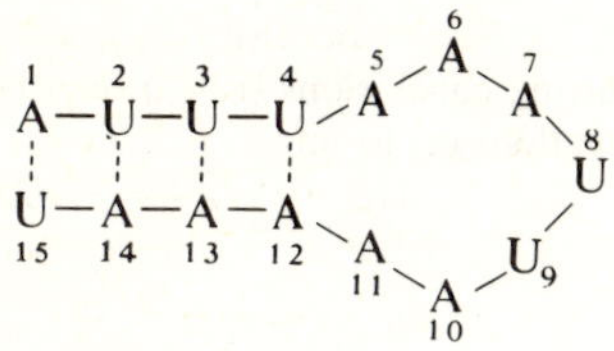

Figure 50 Hairpin loop; ···· depicts H bonding.

(b) There is then a total of *eight* intrachain H bonds, two per base pair.

(c) No, probably not. There are only four base pairs, which means less than a half-turn of double helix (see p. 23).

ITQ 11 The λ_{max} for Tyr = 278 nm; the λ_{max} for Trp = 280 nm, with a shoulder at 289 nm.

ITQ 12 (a) From equation 2.1,

$$A = \epsilon c l, \quad \text{or} \quad c = A/\epsilon l$$

Substituting for A, ϵ and l,

$$c = \frac{0.8}{5.1 \times 10^4 \times 1} \, \text{mol l}^{-1}$$
$$= 1.6 \times 10^{-5} \, \text{mol l}^{-1} \text{ (approx.)}$$

Therefore, the concentration of pepsin in the unknown solution with absorbance 0.8 is 1.6×10^{-5} mol l^{-1}.

(b) Since ϵ, the absorbance of a one molar solution, is given as 5.1×10^4, we can say that the A_{278} of a 35 000 mg ml^{-1} solution is 5.1×10^4.

Therefore the A_{278} of a 10 mg ml^{-1} solution is

$$5.1 \times 10^4 \times 10/35\,000$$

i.e. approximately 14.

ITQ 13 Nucleic acid. Molar absorbance is the absorbance per molecule, and in nucleic acids all the constituent building blocks will contribute, since all five nucleotide bases are UV-absorbing. In proteins only three of the 22 amino acids are UV-absorbing. On average, proteins absorb about twenty times less strongly than nucleic acids in the 'useful' 260–280 nm wavelength range.

ITQ 14 Any change in absorbance of the protein at around 400 nm can be unambiguously assigned to a change in tyrosine environment. Unmodified tyrosine absorbs in the UV, as do all the other chromophores (tryptophan, phenylalanine and the peptide bond), but there is no naturally occurring amino acid that absorbs in the visible region.

SAQ answers

SAQ 1 (a) Three.

(b) Six. Pitch $p = n \times d$.

(c) Two. These can be seen lying parallel to the helix axis.

(d) Weaker, because the three combining atoms N—H—O do not lie in a straight line as they do in the α-helix.

SAQ 2 (a) False. The chain folds spontaneously to the conformation of *minimum* free energy.

(b) True. This arrangement makes for minimum free energy by allowing internal hydrophobic residues to form van der Waals bonds with each other. Further bond energy comes from H and ionic bonds formed between polar residues—now on the outside—and the surrounding H$_2$O molecules.

SAQ 3 (a) 75 EU. In the denaturing urea solution, the protein unfolds to a random coil conformation with virtually no catalytic activity (step 2). Careful removal of urea regenerates a substantial proportion of the enzyme in the active conformation.

(b) 2 EU. Chemical modification causes a covalent change in primary structure. Since primary→tertiary, this modified protein is unlikely to refold to an active conformation. (See Unit 4 for a similar experiment on ribonuclease.)

SAQ 4 Seven. The third one down is partially obscured by an α-helix.

SAQ 5 (a) About $4\frac{1}{2}$ turns. (b) Sixteen. An α-helix has 3.6 residues per turn. This carboxypeptidase helix has approximately $4\frac{1}{2}$ turns, hence around 16.2 (i.e. 4.5×3.6) residues.

SAQ 6 β-pleated sheet. Note how the H bonds (in red) run at right angles to the main chain, rather than parallel to it as in an α-helix.

SAQ 7 Interpretation (b) is correct, since sedimentation results suggest that Y is *smaller* than X. A mutation is any change in primary structure of a protein. This may be removal or insertion of amino acid residues, or substitution of one for another, e.g. Arg changes to His. In the example here the mutation is probably in the subunit interaction site rather than the active site, since enzyme Y is not only less active than X but considerably smaller than it. This can be seen from the S values (see Unit 1, Section 1.7.1).

SAQ 8 The plant DNA. This requires more energy for denaturation (higher T_m value) because of its high proportion of triple H-bonded base pairs (G+C) rather than double H-bonded pairs (A+T) (see Figure 15).

SAQ 9 Adenine–thymine base pairs. These are held together by only two H bonds per base pair and are therefore more readily separated than guanine–cytosine base pairs which have three H bonds per base pair.

SAQ 10 The nucleotide chain will fold as shown in Figure 51. In this way the anticodon loop is stabilized by six H-bonded pairs along the stem, while the anticodon sequence CUA (complementary to the Asp codon, GAU) is prominently exposed.

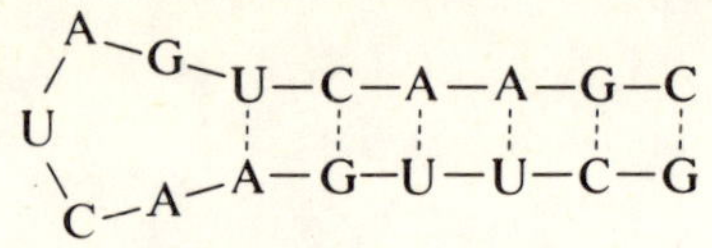

```
      A  G
     /   \  U—C—A—A—G—C
   U       |  |  |  |  |  |
     \   A  A—G—U—U—C—G
      C
```

Figure 51 Secondary structure of imaginary anticodon loop in tRNA; ···· depicts H bonding.

SAQ 11 (a) UV absorption or ORD. (X-ray crystallography and NMR are laborious methods giving higher resolution than is required here; chemical modification gives information only about specific residues; fluorescence requires tryptophan or tyrosine residues.)

(b) UV absorption or ORD.

(c) Fluorescence emission spectroscopy, a particularly sensitive technique.

(d) UV absorption, ORD or fluorescence.

Unfolding and refolding involve widespread changes in conformation and are best studied by methods sensitive to interactions between different parts of the molecule, particularly between chromophoric residues.

SAQ 12 Suitable techniques would be: fluorescence; UV absorption, particularly where the interaction involves changes in the region of tyrosine chromophores; NMR if the enzyme is both plentiful and heat stable. (Chemical modification experiments like those described in the text may also be used, to give an idea of whether or not a specific residue lies in the enzyme active site.)

Substrate–enzyme interactions produce only small changes in conformation of the enzyme and are therefore best studied by looking at individual residues in the macromolecule, rather than gross unfolding phenomena affecting the whole protein. Thus ORD is not appropriate.

SAQ 13 (a) ORD, UV, (NMR), chemical modification, fluorescence.

(b) All the techniques.

(c) As (a).

(d) All the techniques.

A high resolution method is clearly always desirable, but this should in each case be substantiated by data from other methods. NMR is mentioned only in parentheses, to emphasize that the high resolution potential of this technique has very seldom yet been realized because of the assignment problem mentioned in the text. Chemical modification is a good back-up method giving information on the inside versus outside location of specific residues.

SAQ 14 (a) Conclusion (ii) is the only interpretation possible with this limited amount of data. As you can see from Figure 31 in the text, both tyrosine and tryptophan absorb in this part of the UV spectrum, and so the change at 295 nm could be caused by conformational changes in the region of either residue.

(b) Conclusion (i). Nitrotyrosine is yellow, i.e. it absorbs in the visible region of the spectrum, and in this way can be observed independently of the other UV-absorbing chromophores. The ambiguity about Tyr or Trp residues is thus resolved.

(c) Conclusion (i). Again, chemical modification allows Tyr to be distinguished from Trp. In this case, the conformational change is followed not spectroscopically but by amino acid analysis of the excised active site peptide.

Acknowledgements

Grateful acknowledgement is made to the following for material used in this unit:

Figure 4 from R. E. Dickerson and I. Geis (1969) *The Structure and Action of Proteins*, Harper and Row; *Figure 11* from Professor David Phillips; *Figure 14* from C. B. Anfinsen (1973) Principles that govern the folding of protein chains *Science N.Y.* **181**, copyright © American Association for the Advancement of Science; *Figure 16* from V. Bloomfield and R. E. Harrington (1974) *Biophysical Chemistry*, W. H. Freeman and Co., first published in *Scientific American*, October 1954, by permission of Paul Weller, photographer; *Figure 18* from H. R. Mahler and E. H. Cordes (1971) *Biological Chemistry*, 2nd edn, Harper and Row; *Figure 21* from A. Rich *et al.* (1973) 3-D structure of yeast phe tRNA, *Science, N.Y.*, **179**, copyright © American Association for the Advancement of Science; *Figure 22* from J. Ladner *et al.* (1975) Structure of yeast phe tRNA, *Proc. Nat. Acad. Sci.*, **12**, National Academy of Science; *Figures 23(a) and (b)* from J. Robertus *et al.* (1974) Structure of yeast phe tRNA, *Nature, Lond.*, **250**, Macmillan and author; *Figure 27* from M. Rhoades and C. A. Thomas, Jr. (1968) *J. Mol. Biol.*, **37**, 41; *Figure 28* from M. J. Waring's (1968) article in *Nature, Lond.*, **219**, Macmillan and author; *Figure 29* from W. T. Perry *et al.* (1973) A phosphorylated light chain of myosin, *Biochemical Journal*, **135**, The Biochemical Society; *Figure 32* from L. N. Johnson and R. Wolfenden (1970) Changes in absorption structure, *Journal of Molecular Biology*, **47**, Academic Press and author; *Figures 33(b) and (c)* from D. B. Wetlaufer (1962) UV spectra of proteins, *Advances in Protein Chemistry*, **17**, Academic Press and author; *Figures 34 and 41* from A. J. Furth and D. B. Hope (1970) Studies on chemical modification, *Biochemical Journal*, **116**, The Biochemical Society and authors; *Figure 35* from V. A. Bloomfield *et al.* (1974) *Physical Chemistry of Nucleic Acids*, Harper and Row; *Figure 37* from R. Marsh *et al.* (1955) An investigation of the structure of fibroin, *Biochimica et Biophysica Acta*, **16**; *Figure 38* from H. Neurath (1964) *The Proteins*, 2nd edn, vol. II, Academic Press and M. H. F. Wilkins; *Figures 39(a) and (b)* from H. Neurath (1964) *The Proteins*, 2nd edn, vol. II, Academic Press and R. E. Dickerson.

S322 BIOCHEMISTRY AND MOLECULAR BIOLOGY

LIVERPOOL INSTITUTE OF
HIGHER EDUCATION
THE MARKLAND LIBRARY